T0252765

Programming Multi-Agent Systems
in AgentSpeak using *Jason*

Wiley Series in Agent Technology

Series Editor: Michael Wooldridge, *University of Liverpool, UK*

The 'Wiley Series in Agent Technology' is a series of comprehensive practical guides and cutting-edge research titles on new developments in agent technologies. The series focuses on all aspects of developing agent-based applications, drawing from the Internet, Telecommunications, and Artificial Intelligence communities with a strong applications/technologies focus.

The books will provide timely, accurate and reliable information about the state of the art to researchers and developers in the Telecommunications and Computing sectors.

Titles in the series:
Padgham/Winikoff: Developing Intelligent Agent Systems 0470861207 (June 2004)
Bellifemine/Caire/Greenwood: Developing Multi-Agent Systems with JADE 978-0-470-05747-6 (February 2007)

Programming Multi-Agent Systems in AgentSpeak using *Jason*

Rafael H. Bordini

University of Durham, UK

Jomi Fred Hübner

University of Blumenau, Brazil

Michael Wooldridge

University of Liverpool, UK

Other Wiley Editorial Offices

John Wiley & Sons Inc., 111 River Street, Hoboken, NJ 07030, USA

Jossey-Bass, 989 Market Street, San Francisco, CA 94103-1741, USA

Wiley-VCH Verlag GmbH, Boschstr. 12, D-69469 Weinheim, Germany

John Wiley & Sons Australia Ltd, 42 McDougall Street, Milton, Queensland 4064, Australia

John Wiley & Sons (Asia) Pte Ltd, 2 Clementi Loop #02-01, Jin Xing Distripark, Singapore 129809

John Wiley & Sons Canada Ltd, 6045 Freemont Blvd, Mississauga, Ontario, L5R 4J3, Canada

Wiley also publishes its books in a variety of electronic formats. Some content that appears
in print may not be available in electronic books.

Anniversary Logo Design: Richard J. Pacifico

Library of Congress Cataloging-in-Publication Data

Bordini, Rafael H.
    Programming multi-agent systems in AgentSpeak using Jason / Rafael H.
Bordini, Jomi Fred HŸbner, Michael Wooldridge.
        p. cm.
    Includes bibliographical references.
    ISBN 978-0-470-02900-8 (cloth)
    1.   Intelligent agents (Computer software) 2.   Computer programming.
I. Hübner, Jomi Fred. II. Wooldridge, Michael J., 1966- III. Title.
    QA76.76.I58B67 2007
    006.3'3 — dc22
                                                        2007021099
British Library Cataloguing in Publication Data

A catalogue record for this book is available from the British Library

ISBN   978-0-470-02900-8 (HB)

Typeset in 11/13.6pt Garamond by Laserwords Private Limited, Chennai, India
Printed and bound in Great Britain by Antony Rowe Ltd, Chippenham, Wiltshire
This book is printed on acid-free paper responsibly manufactured from sustainable forestry
in which at least two trees are planted for each one used for paper production.

To Idahyr (*in memoriam*), Maria, Ricardo, Lizete, Roberto, Renato, Rubens, Liliane and Thays.   (RHB)

To Ilze, Morgana and Thales.   (JFH)

To Lily May and Thomas Llewelyn.   (MW)

Contents

Preface

A typical day:

> You are waiting for the bus to take you to the airport: it is late, and you begin to be concerned about whether, if it is much later, you will have enough time to make your flight. You decide to abandon your plan of travelling by bus, and flag down a taxi instead. The taxi driver does not know which terminal your flight departs from; neither do you. As you approach the airport, you telephone the information line number printed on your ticket, and you are told that the flight departs from terminal one. The taxi drops you off at terminal one, and you enter the terminal, to be confronted by long queues at the check-in counters. You realise that, since you have no check-in baggage, you can use an express check-in counter, and within a minute, an attendant is checking you in. The attendant asks you whether you packed the bag yourself. No, you reply, but immediately point out to her that the bag was packed by your partner. You are given a boarding card and told to proceed through security; you do not know where the security gate is, so you ask. Proceeding through security, the metal detector goes off; you realise your mobile phone is in your pocket, so you remove it, but the detector still goes off. You remove your belt, then your wallet, and you are allowed through. In the departure lounge, you look for a cafe, and buy a sandwich. While you are eating your sandwich, you open your laptop, and quickly check through your email, deleting a dozen or so spam messages, scanning the genuine mail for anything important. While doing this, you keep half an eye on the departure board; your flight is called, so you finish your sandwich, shutdown your laptop, and board.

There does not seem to be anything magical about this scenario, does there? Yet an enormous range of human skills are in evidence here, which are worth identifying. Here are some examples:

- We start the day with some goal in mind (be on a particular flight at the airport at a particular time), and we have some idea of how we will achieve this – we have a plan of action to get us to the airport, involving a bus. Yet the plan is only partially elaborated – we do not know, for example, exactly what time the bus will turn up, exactly where it will drop us, how much it will cost, or where exactly at the airport the check-in desks or security gate are. However, we are able to carry out such thinly specified plans, filling in the details as required on the fly.

- When things go wrong, as they do in this scenario, we seem to be able to recover more-or-less seamlessly. The bus does not turn up: but we do not do a system crash, and we do not give up and go home. We quickly develop an alternative plan (flag down a taxi), and carry it out.

- While we are executing our plan we realise that we need some information – the terminal number – for our current plan to be successful. We perform an action (calling the airport information number) that will furnish us with this information.

- Throughout the day, we cooperate and coordinate our activities with other people – asking questions, answering their questions, solving problems together. We are able to anticipate their needs, and act to accommodate them. When the taxi driver tells us that she does not know which terminal our flight departs from, we work with her to find out. When the check-in attendant asks us if we packed the bag ourselves and we answer 'no', we anticipate that this will lead to further questions, and so we anticipate these, pro-actively providing the information we know the attendant wants. When we are going through the metal detector, we work with the security personnel to find out what is setting off the alarm.

- We interleave multiple activities, concurrently, each of them attempting to achieve one of our goals. We eat our sandwich, while reading email, while watching the departure board for details of our flight departure.

Still, most people would not regard this scenario as anything magic: it is everyday; positively routine, hardly worth mentioning. For researchers in artificial intelligence, however, the scenario is positively dripping with magic. The kinds of skills we describe above, which seem so everyday to us, are notoriously, frustratingly, irritatingly hard to deploy in computer systems. We do not get computers to do things for us by giving them some high-level goal, and simply letting them get on with it: we have to tell them what to do by giving them a precise, tediously detailed list of instructions, called a 'program', which the computer blindly executes. When computer systems encounter unexpected scenarios, and things do not go as we anticipated, they do not seamlessly recover and develop alternative courses of action. We get an uncaught exception at best; a complete system failure at worst. Computers do not cooperate with us, anticipate our needs, and coordinate their activities around us. Indeed, computers often seem sullen, uncooperative and indeed positively unhelpful. Nor can we communicate them in any high-level way: everything that they do seems to have to be communicated to them by us selecting items from menus, clicking on a button or dragging something across a window – the idea that the computer anticipates our needs is definitely laughable.

This book is all about writing computer programs that have some flavour of the kind of skills illustrated above. We should put in an important disclaimer at this point: we do not claim that our programs come close to the abilities evident in the scenario. However, after you have finished reading this book, you should have a reasonable understanding of what the main issues are, and what sorts of techniques are currently available for programming software with these kinds of skills.

The book describes a programming language called 'AgentSpeak', and specifically, an implementation of the AgentSpeak language called *Jason*. The basic idea behind programming in AgentSpeak is to define the *know-how* of a program, in the form of *plans*. By 'know-how', we mean *knowledge about how to do things*. In the above scenario, for example, our know-how relates to travelling to the airport, and for this we have two plans: one involving a bus, the other involving a taxi. Programming in AgentSpeak thus involves encoding such plans, and AgentSpeak provides a rich, high-level language that allows programmers to capture this 'procedural knowledge' in a transparent manner. Plans in AgentSpeak are also used to characterise *responses* to events. In this way, we can build programs that not only systematically try to accomplish goals by making use of the know-how we provide, but they can also be responsive to their changing environment. The resulting programming paradigm incorporates aspects of conventional, so-called *procedural* programming, as embodied in languages like Pascal and C, as well as a kind of *declarative*, *deductive* style of programming, as embodied in Prolog. Ultimately, though, while these analogies may be useful, the style of programming in AgentSpeak is fundamentally different from either of these paradigms, as you will see in the remainder of the book. These differences are summed up in the following way: we refer to computer programs in AgentSpeak as *agents*. This is because they are *active* pieces of software, not dumbly providing services for us or other pieces of software, but constantly trying to achieve goals in their environment, making use of their know-how.

Another important aspect of AgentSpeak and *Jason* in particular is that it is designed with cooperation in mind. Thus, *Jason* comes with a rich environment that will enable agents to communicate and coordinate with one another in a high-level way. Communication in *Jason* is also rather different from the communication you may be familiar with from other programming languages. It is not concerned with low-level issues (getting a byte from *A* to *B*); nor is it concerned with service provision via methods or the like. The focus is on *knowledge level* communication: it addresses issues such as how an agent can communicate its beliefs to another, how an agent can delegate one of its goals, and how an agent can communicate its own know-how.

The book you are holding provides a practical introduction to AgentSpeak and *Jason*; it gives a detailed overview of how to make the most of the *Jason* language, hints and tips on good AgentSpeak programming style, a detailed overview of the *Jason* programming environment, a formal semantics for the language, and some pointers to where the language came from, as well as where it is going. Whether you are reading this book in order to build agent systems, or just curious about a new style of programming, we hope you will find it instructive, useful and thought provoking. We conclude with advice that is by now traditional for programming books. While you might understand how the language works in principle from reading the book, there is ultimately no substitute for having a go yourself. *Jason* is freely available: so why not download it, try, learn and enjoy. Links to the software download page and teaching resources (slides, solved exercises, etc.) are available at `http://jason.sf.net/jBook`.

Structure of this Book

- The first two chapters of the book provide essentially background material, giving an introduction to agents and multi-agent systems, and a short overview of the BDI model of agency, which underpins AgentSpeak. If you know something about agents, and want to go straight into the language, then you can skip these chapters.

- Chapter 3 is probably the most important chapter in the book: it gives a detailed introduction to the AgentSpeak language as implemented in the *Jason* system. If you want to know how to program in AgentSpeak, you need to read this chapter.

- Chapter 4 describes how the *Jason* interpreter works: if you want to understand exactly how your programs will behave, and optimise your code, then you need to read this chapter to understand how the interpreter deals with them.

- Chapter 5 describes *Jason*'s implementation of communication.

- Chapter 6 describes the support available for developing (in Java) simulated environments where agents are to be situated.

- Chapter 7 shows how you can replace parts of *Jason* with your own Java code. If you want your agent to be embedded in some system, or to talk with some legacy system, for example, then you need to read this chapter.

- Chapter 8 describes various 'programming patterns' that can be used when programming with AgentSpeak using *Jason*. Roughly, this chapter is about programming *style*.

- Chapter 9 gives some working case studies.

- Chapter 10 presents formal semantics for AgentSpeak.

- Chapter 11 presents some conclusions, and ongoing research issues.

Thus, if you just want to understand the basic principles of AgentSpeak, then Chapters 3 and 4 are the most important. If you want to do serious system development, then Chapters 3 – 9 are the most important. If you want to do research on AgentSpeak itself, then Chapters 10 and 11 are also important.

Acknowledgements

A significant body of research has helped to shape *Jason* in its current form. Such research was done by the authors of this book in collaboration with many colleagues, including: Michael Fisher, Willem Visser, Álvaro Moreira, Renata Vieira, Natasha Alechina, Brian Logan, Viviana Mascardi, Davide Ancona, Antônio Carlos da Rocha Costa and Fabio Okuyama. The programming of the *Jason* platform was done by Jomi Hübner and Rafael Bordini with much appreciated support from Joyce Martins, who helped us improve the grammar used for the parsing of AgentSpeak code. With the many extensions to the language that we did over the last few years, Joyce has been a continued source of help: many thanks to her! Many students have helped in various different ways, and we would like to thank specially some of the first students to be involved with implementations of AgentSpeak interpreters and using them, for their brave work: especially Rodrigo Machado, but also Daniel Basso, Rafael Jannone and Denise de Oliveira. The undergraduate students in the Multi-Agent Systems module at Durham University and final year project students at Durham and FURB using *Jason* also contributed enormously in this development, including Alisson Appio, Daniel Dalcastagne, George Millard, Marios Richards, Daniel Tallentire, Karlyson Vargas and Daniel Wilkinson. Equally, students in various courses/projects in Brazil, Netherlands, England, Italy, Portugal and Australia, as well as subscribers to *Jason* email lists, should be thanked for the courage in using *Jason* during a time it was evolving very quickly; their experience certainly helped us a lot. Many others have also helped us, and the *Jason* manual lists some more names; many thanks to all those who helped (e.g. by pointing out bugs). Finally, we would like to thank Patricia Shaw and Berndt Farwer who read drafts of some chapters of this book, and of course Birgit, Richard and Sarah from Wiley for all the help during the (longer than hoped) period it took us to write this book.

1

Introduction

This book is all about a computer programming language called AgentSpeak, and a particular implementation of AgentSpeak called *Jason*. The AgentSpeak language is intended for developing *multi-agent* systems. Before we start to investigate how to program with AgentSpeak, it seems appropriate to try to understand in more detail what multi-agent systems are, and some of the ideas that underpin the language.

1.1 Autonomous Agents

To better understand what we mean by the terms 'agent' and 'multi-agent systems', let us consider how agents relate to other types of software. Start by considering *functional* programs, which are possibly the simplest type of software from the point of view of software development and software engineering. A functional program takes some input, chews over this input, and then on the basis of this, produces some output and halts. A compiler is an example of such a program: the input is some source code (e.g. a `.java` file), and the output is bytecode (`.class` files), object code or machine code. When we learn how to program, the types of program we typically construct are of this type: the sorts of exercises we set to programmers in an introductory Java class are things like 'read a list of numbers and print the average'. Functional programs are so called because, mathematically, we can think of them as functions $f : I \rightarrow O$ from some domain I of possible inputs (source code programs, in our 'compiler' example) to some range O of possible outputs (bytecode, object code, etc). We have a range of well-established techniques for developing such programs; the point is that, from the standpoint of software development, they are typically straightforward to engineer.

Programming Multi-Agent Systems in AgentSpeak using Jason   R.H. Bordini, J.F. Hübner, M. Wooldridge
© 2007 John Wiley & Sons, Ltd

Unfortunately, many programs do not have this simple input – compute – output operational structure. In particular, many of the systems we need to build in practice have a 'reactive' flavour, in the sense that they have to *maintain a long-term, ongoing interaction with their environment*; they do not simply compute some function of an input and then terminate:

> Reactive systems are systems that cannot adequately be described by the *relational* or *functional* view. The relational view regards programs as functions . . . from an initial state to a terminal state. Typically, the main role of reactive systems is to maintain an interaction with their environment, and therefore must be described (and specified) in terms of their on-going behavior . . . [E]very concurrent system . . . must be studied by behavioral means. This is because each individual module in a concurrent system is a reactive subsystem, interacting with its own environment which consists of the other modules. [77]

Examples of such programs include computer operating systems, process control systems, online banking systems, web servers, and the like. It is, sadly, a well-known fact that, from the software development point of view, such *reactive* systems are *much* harder to correctly and efficiently engineer than functional systems.

A still more complex class of systems is a subset of reactive systems that we will call *agents*. An agent is a reactive system that exhibits some degree of *autonomy* in the sense that we delegate some task to it, and the system itself determines how best to achieve this task. We call such systems 'agents' because we think of them as being active, purposeful producers of actions: they are sent out into their environment to achieve goals for us, and we want them to actively pursue these goals, figuring out for themselves how best to accomplish these goals, rather than having to be told in low-level detail how to do it. We can imagine such agents being delegated a task like booking a holiday for us, or bidding on our behalf in an online auction, or cleaning our office space for us, if they are robotic agents.

1.2   Characteristics of Agents

Let us try to be a little more precise about what sorts of properties we are thinking of when we talk about agents. We consider agents to be systems that are *situated* in some *environment*. By this, we mean that agents are capable of *sensing* their environment (via *sensors*), and have a repertoire of possible *actions* that they can perform (via *effectors* or *actuators*) in order to modify their environment. The key question facing the agent is how to go from sensor input to action output: how to *decide what to do* based on the information obtained via sensors. This leads to the

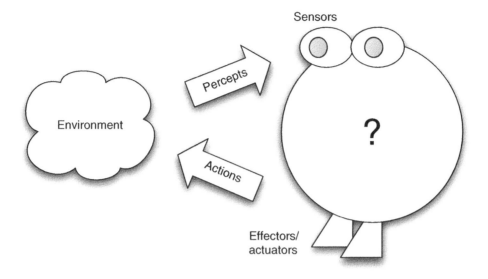

Figure 1.1 Agent and environment (after [82, p. 32]).

view of an agent as shown in Figure 1.1. As we will see, in AgentSpeak, deciding what to do is achieved by manipulating *plans*.

The environment that an agent occupies may be physical (in the case of robots inhabiting the physical world) or a software environment (in the case of a software agent inhabiting a computer operating system or network). We think of decisions about what action to perform being translated into actual actions via some mechanism external to the agent; usually, this is achieved via some sort of API. In almost all realistic applications, agents have at best *partial* control over their environment. Thus, while they can perform actions that change their environment, they cannot in general completely control it. Very often this is because there will be other agents in the environment, who exhibit control over their part of the environment.

Apart from being situated in an environment, what other properties do we expect a rational agent to have? Wooldridge and Jennings [104] argued that agents should have the following properties:

- autonomy;
- proactiveness;
- reactivity; and
- social ability.

Autonomy

It is important to realise that autonomy is a very broad spectrum. At one end of the spectrum, we have computer programs such as conventional word processors and spreadsheets, which exhibit little or no autonomy. Everything that happens with such an application happens because you make it happen – you select a menu item, or click on an icon, for example. Such programs, by and large, do not *take the initiative* in any sense. At the other end of the autonomy spectrum are you and us. You are completely autonomous. You can ultimately choose to believe what you want, and do what you want – although society typically constrains your autonomy in various ways, preventing you from doing certain things, for the sake of you and your peers. You have your own goals, your own agenda, and autonomy means they really are yours: nobody and nothing explicitly dictates them to you. (Of course, you might argue that society tries to shape our beliefs and goals, but that is another story.) In this book, we are interested in computer programs that lie somewhere between these two extremes. Roughly speaking, we want to be able to *delegate* goals to agents, which then decide how best to act in order to achieve these goals. Thus, our agent's ability to construct goals is ultimately bounded by the goals that we delegate. Moreover, the way in which our agents will act to accomplish their goals will be bounded by the *plans* which we give to an agent, which define the ways in which an agent can act to achieve goals and sub-goals. One of the key ideas in AgentSpeak is that of an agent putting together these plans on the fly in order to construct more complex overall plans to achieve our goals.

At its simplest, then, autonomy means nothing more than being able to operate independently in order to achieve the goals we delegate to an agent. Thus, at the very least, an autonomous agent makes independent decisions about how to achieve its delegated goals – its decisions (and hence its actions) are under its own control, and are not driven by others.

Proactiveness

Proactiveness means being able to exhibit *goal-directed* behaviour. If an agent has been delegated a particular goal, then we expect the agent to try to *achieve* this goal. Proactiveness rules out entirely passive agents, who never try to do anything. Thus, we do not usually think of an object, in the sense of Java, as being an agent: such an object is essentially passive until something invokes a method on it, i.e. tells it what to do. Similar comments apply to web services.

Reactiveness

Being *reactive* means being *responsive* to changes in the environment. In everyday life, plans rarely run smoothly. They are frequently thwarted, accidentally or deliberately. When we become aware that our plans have gone wrong, we respond, choosing an alternative course of action. Some of these responses are at the level of 'reflexes' – you feel your hand burning, so you pull it away from the fire. However, some responses require more deliberation – the bus has not turned up, so how am I going to get to the airport? Designing a system which simply responds to environmental stimuli in a reflexive way is not hard – we can implement such a system as a lookup table, which simply maps environment states directly to actions. Similarly, developing a purely goal-driven system is not hard. (After all, this is ultimately what conventional computer programs are: they are just pieces of code designed to achieve certain goals.) However, implementing a system that achieves an effective *balance* between goal-directed and reactive behaviour turns out to be hard. This is one of the key design objectives of AgentSpeak.

Social Ability

Every day, millions of computers across the world routinely exchange information with humans and other computers. In this sense, building computer systems that have some kind of social ability is not hard. However, the ability to exchange bytes is not social ability in the sense that we mean it. We are talking about the ability of agents to *cooperate* and *coordinate* activities with other agents, in order to accomplish our goals. As we will see later, in order to realise this kind of social ability, it is useful to have agents that can communicate not just in terms of exchanging bytes or by invoking methods on one another, but that can communicate at the *knowledge level*. That is, we want agents to be able to communicate their *beliefs*, *goals* and *plans* to one another.

1.3 Multi-Agent Systems

So far, we have talked about agents occupying an environment in isolation. In practice, 'single agent systems' are rare. The more common case is for agents to inhabit an environment which contains *other* agents, giving a *multi-agent system* [103]. Figure 1.2 gives an overview of a multi-agent system. At the bottom of the figure, we see the shared environment that the agents occupy; each agent has a 'sphere of influence' in this environment, i.e. a portion of the environment that they are able to control or partially control. It may be that an agent has the unique ability to control part of its environment, but more generally, and more problematically, we have the possibility that the spheres of influence overlap: that is, the environment is *jointly* controlled. This makes life for our agents more complicated, because to achieve an outcome in the environment that our agent desires, it will have to take into account how the other agents with some control are likely to act.

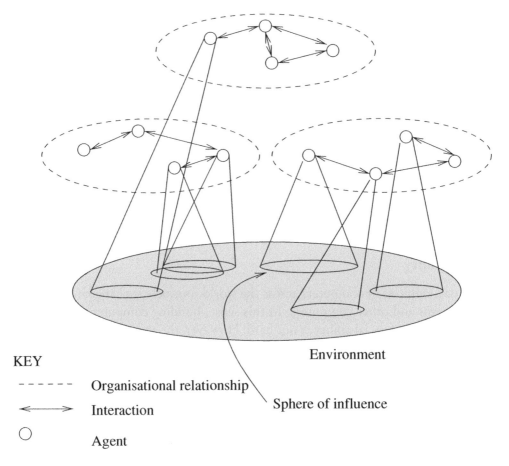

KEY

– – – – – –   Organisational relationship

←——————→   Interaction

○   Agent

Environment

Sphere of influence

Figure 1.2 Typical structure of a multi-agent system (after [58]).

Above the environment, we see the agents themselves, which stand in various organisational relationships to one another (for example, one agent may be the peer of another, or may have line authority over another). Finally, these agents will have some knowledge of each other, though it may be the case that an agent does not have complete knowledge of the other agents in the system.

Programming Languages for Agents and Multi-Agent Systems

We now have some idea of what kinds of properties we are thinking of in our agents. So, suppose we want to program these things: what do these properties tell us about the kinds of programming language or environment that we might use for programming autonomous agents? We can identify the following requirements:

- The language should support delegation at the level of goals. As we noted earlier, when we delegate a task to an agent, we do not generally want to do this by giving the agent an executable description of what to do. Rather, we want to communicate with it at the level of goals: we should be able to describe our goals to an agent in a high-level way, independent of approaches to achieving these goals.

- The language should provide support for goal-directed problem solving. We want our agents to be able to act to achieve our delegated goals, systematically trying to achieve them.

- The language should lend itself to the production of systems that are responsive to their environment.

- The language should cleanly integrate goal-directed and responsive behaviour.

- The language should support knowledge-level communication and cooperation.

These are the main requirements that AgentSpeak and *Jason* are intended to fulfill. In the following section, we will give a very brief introduction to AgentSpeak, which will give a feel for how some of these features are provided.

1.4 Hello World!

When introducing a new programming language, it has become the tradition to give a short example program, the purpose of which is simply to display the text 'Hello World!' to the programmer.[1] For example, here is a 'Hello World' program in Java:

```
public class HelloWorld {
   public static void main( String args[] ) {
      System.out.println( "Hello World!" );
   }
}
```

Trivial though they are, running a 'Hello World' program helps to give a programmer confidence with the new language, and very often they give a useful insight

---

[1]There are even web sites devoted to 'Hello World' programs: http://www.roesler-ac.de/wolfram/hello.htm is one example, with 'Hello World' programs for hundreds of languages.

into the 'mind set' of the language. Therefore, we strongly encourage you to try this exercise:

```
started.

+started <- .print("Hello World!").
```

Let us try to understand a little of what is going on here, although we will save the details for later.

The first thing to understand is that this constitutes the definition of a single agent. This definition will often be saved in a single file, and let us suppose that on our system we have called this file `hello.asl`; the `.asl` extension will be used for all our AgentSpeak programs. Now, the definition of an agent in AgentSpeak consists of two parts:

- the agent's *initial beliefs* (and possibly *initial goals*); and

- the agent's *plans*.

The first line defines one initial belief for our agent. (Although there is only one initial belief here, we could have given a list.) AgentSpeak does not have variables as in programming languages such as Java or C; the constructs are specific agent notions, such as beliefs, goals and plans. We need 'beliefs' because the intuition is that they represent the information that the agent has currently been able to obtain about its environment. The full stop, '.', is a syntactic separator, much as a semicolon is in Java or C. Therefore, when our agent first starts running, it will have the single belief `started`; intuitively, it has the belief that it has started running. Notice that there is no magic in the use of the term 'started'; we could just as well have written the program as follows, and we would have obtained exactly the same result:

```
cogitoErgoSum.

+cogitoErgoSum <- .print("Hello World!").
```

So, while its important and helpful for programmers to pick sensible and useful names for beliefs, the computer, of course, could not care less. One rule worth remembering at this stage is that beliefs must start with a lowercase letter.

Next, we have the line

```
+started <- .print("Hello World!").
```

which defines a plan for the agent – which is in fact the only plan this agent has. Intuitively, we can read this plan as meaning whenever you come to believe 'started', print the text 'Hello World!'.

The plan, like all AgentSpeak plans, comes in three parts. The first part is a *triggering event*. In this case, the triggering event is simply

```
+started
```

The symbol '+' in this context means 'when you acquire the belief . . .', and so overall the triggering condition is 'when you acquire the belief "started"'. We know from the above discussion that the agent acquires this belief when it starts executing, and so in sum, this plan will be triggered when the agent starts executing.

However, what does it mean, to trigger a plan? The idea is that the trigger of a plan defines the *events* that it is useful for handling. In this case, the event is the acquisition of a particular new belief. A plan is triggered when events occur which match its trigger condition, and when this happens, the plan becomes 'active'; it becomes something that the agent 'considers doing'. However, before an agent selects a plan to become active, it checks that the *context* of the plan is appropriate. In the hello world case, the context is in fact empty, which can be understood as meaning 'this plan is always good'. As we will see later, in the context part of plans, we can define complex conditions which an agent uses to determine whether or not to choose a particular plan for a given event.In particular, an agent can have *multiple* plans triggered by the *same* event which deal with this event in *different* ways: thus an agent can have multiple different responses to events, and can choose between these depending on the situation in which it currently finds itself.

In this case, there is just one plan that can be triggered by the event, and since the context is empty, it is always applicable, and so the AgentSpeak interpreter *directly executes* the body of that plan. In this case, the body of the plan is very simple, containing a single action:

```
.print("Hello World!").
```

As you might guess, the effect of this action is simply to display the text 'Hello World!' on the user's console. Running this example on *Jason*, the result is that the following gets displayed in the user's console (see Figure 1.3):

```
[hello] saying: Hello World!
```

There are several points to note here. First, although .print(...) looks like a belief, it is in fact an action, as it appears in the plan body; to give the reader some syntactic clues when reading a *Jason* program, 'internal' actions begin with a full stop (actions normally change the environment, but not *internal* actions). In fact, .print(...) is a pre-defined internal action in *Jason*: other pre-defined actions

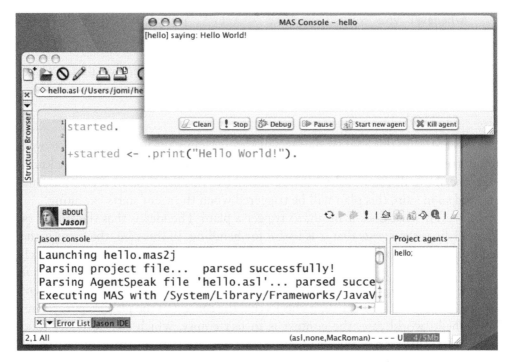

Figure 1.3 Running the 'Hello World!' program with *Jason*.

include, for example .send(...) and .broadcast(...) for agent communication, and .stopMAS, to cause a halt to the execution of the multi-agent system. Appendix A describes other internal actions available in *Jason*.

Don't be misled by the very simple structure of this example into thinking that plan bodies are just like sub-routines, methods or procedures from conventional programming languages: they are in fact very much more than that. For example, one of the key ideas in AgentSpeak is that we can have a *goal* in a plan body. The idea is that, when the interpreter comes to a goal in a plan body, it tries to find a plan that achieves this goal; as we discussed above, there may be several such plans, which may or may not match the current context. The upshot of this is that it is possible, within AgentSpeak, to invoke code *with reference to the effect of the code*, and the same invocation, in different contexts, results in different code being invoked. This is a very substantial difference to conventional languages like Java or C.

Finally – and this may at first sight seem somewhat perverse to those with a grounding in the theory of computing – if you run this example, you will see that *the program does not terminate!* We are used to thinking of non-termination (infinite loops and the like) as a bad thing. In the case of AgentSpeak, however, we are after agents that are *aware of* and *respond to* their environment. It is therefore quite natural that they should not terminate: in fact, if ever the belief `started` was deleted and then added to our agent's belief base again, then the agent would fire into life, once again printing the 'Hello World!' text. Of course, this does not happen in this example; the agent patiently watches for events that never occur. Programmers familiar with the Prolog language might contrast this behaviour with that of Prolog programs: a Prolog program never does anything until a user asks a query, and the behaviour that the program then generates is a side-effect of trying to prove a theorem.

Another Simple Example: Computing the Factorial

To conclude, we will present another short program, which does something marginally more useful than just displaying some text; it is again a classical example rather than a typical agent program such as those shown later in the book. Specifically, the code computes the factorial[2] of 5. The point about the factorial example is that it illustrates how loops' work, and for some languages, which make heavy use of recursion, the factorial example can be very illustrative. We will start with an example which computes the factorial, even though it is not a very elegant example of AgentSpeak code:

```
fact(0,1).

+fact(X,Y)
   :  X < 5
   <- +fact(X+1, (X+1)*Y).

+fact(X,Y)
   :  X == 5
   <- .print("fact 5 == ", Y).
```

As before, the first line defines the agent's initial belief: the belief `fact(0,1)` means that the agent believes the factorial of 0 is 1. In this example, a belief of the form `fact(X,Y)` will mean that the factorial of X is Y. So, when the agent starts executing, this belief is added to the agent's belief base.

We have two plans in this example: they have the same triggering condition, but different contexts. The first plan may be explained as follows:

> whenever you acquire the belief that the factorial of X is Y, where $X <$ 5, add to your beliefs the fact that the factorial of $X + 1$ is $(X + 1) * Y$.

---

[2]If n is a positive integer, then the factorial of n, written $n!$, is $n * (n - 1) * (n - 2) * \cdots * 2 * 1$; for convenience, we define $0! = 1$. We know that *you* know what factorial means, this is for those other people who do not know.

The second plan can be understood as follows:

> whenever you acquire the belief that the factorial of 5 is Y, print out
> the value of Y.

Therefore, when the agent starts executing, and the belief `fact(0,1)` is added
to its belief set, the first plan is triggered; the context condition is satisfied, (since
$0 < 5$), and so the agent then adds the belief `fact(1,1)`; this again triggers the
plan, resulting in the belief `fact(2,2)` being added, the plan again fires, resulting
in `fact(3,6)` being added, and so on. Notice that the second plan will not become
active as a consequence of these events, since the context part (`X == 5`) is not
satisfied. Eventually, the belief `fact(5,120)` is added, and at this point, the context
condition of the first plan is not satisfied, while the context of the second plan is.
As a result, when running this example with *Jason*, we get the following displayed
on the console:

```
[fact] saying: fact 5 == 120
```

This example shows how multiple plans with the same trigger condition but
different contexts can respond to events in different ways.

Unfortunately, this example is rather messy with respect to AgentSpeak programming style. The problem is that it does not make use of recursion; instead, it
fills up the agent's belief base with `fact(0,1)` up to `fact(5,120)` predicates. This
does not matter so much in this case, but in general, this is not an efficient way of
doing things in AgentSpeak. So let us consider a slightly more complex, but much
more efficient, version of the program.

```
!print_fact(5).

+!print_fact(N)
    <- !fact(N,F);
       .print("Factorial of ", N, " is ", F).

+!fact(N,1) : N == 0.

+!fact(N,F) : N > 0
    <- !fact(N-1,F1);
       F = F1 * N.
```

The first line defines the agent's initial goal: the exclamation mark indicates that
this is a goal to be achieved. Therefore, the initial goal of the agent is to print the
factorial of 5. This can be achieved by first calculating the factorial of N and then
printing out the result, as stated in the first plan.

We have two more plans in this example, both for calculating the factorial. The first of them may be explained as follows:

> whenever you acquire the goal to calculate the factorial of 0, this is known to be 1, therefore nothing else needs to be done.

The second plan can be understood as follows:

> whenever you acquire the goal to calculate the factorial of N, provided N > 0, the course of action to take is to first calculate the factorial of N − 1, and then multiply that value by N to get the factorial of N.

When running this example with *Jason*, we get the following displayed on the console:

```
[fact] saying: Factorial of 5 is 120
```

2

The BDI Agent Model

One of the most interesting aspects of AgentSpeak is that it was inspired by and based on a model of human behaviour that was developed by philosophers. This model is called the *belief – desire – intention* (BDI) model. Belief – desire – intention architectures originated in the work of the Rational Agency project at Stanford Research Institute in the mid-1980s. The origins of the model lie in the theory of human practical reasoning developed by the philosopher Michael Bratman [20], which focuses particularly on the role of intentions in practical reasoning. The conceptual framework of the BDI model is described in [22], which also describes a specific BDI agent architecture called IRMA.

Our aim in this chapter is to describe the BDI model of agency, and to give some idea of how it leads naturally to the kind of software architecture and programming model that are embodied by AgentSpeak.

2.1 Agent-Oriented Programming

The first idea that we need to address is a central one in the BDI model: the idea that we can talk about computer programs as if they have a 'mental state'. Thus, when we talk about a belief – desire – intention system, we are talking about computer programs with computational analogues of beliefs, desires and intentions. This may seem like plain old anthropomorphism if you are familiar with conventional programming paradigms, but as we shall see, it can be both useful and legitimate. Roughly, we think of the distinction between beliefs, desires and intentions as follows:

- Beliefs are information the agent has about the world. This information could be out of date or inaccurate, of course. In a conventional computer program, we might have a variable `reactor32Temp`, which is supposed to hold

Programming Multi-Agent Systems in AgentSpeak using Jason   R.H. Bordini, J.F. Hübner, M. Wooldridge
© 2007 John Wiley & Sons, Ltd

the current temperature of reactor 32. We can think of this variable as being one of the programs beliefs about its environment. The idea in AgentSpeak is similar, except that we explicitly think of such representations as beliefs.

- Desires are all the possible states of affairs that the agent *might* like to accomplish. Having a desire, however, does not imply that an agent acts upon it: it is a *potential* influencer of the agent's actions. Note that it is perfectly reasonable for a rational agent to have desires that are mutually incompatible with one another. We often think and talk of desires as being *options* for an agent.

- Intentions are the states of affairs that the agent has decided to work towards. Intentions may be goals that are delegated to the agent, or may result from considering options: we think of an agent looking at its options and choosing between them. Options that are selected in this way become intentions. Therefore, we can imagine our agent starting with some delegated goal, and then considering the possible options that are compatible with this delegated goal; the options that it chooses are then intentions, which the agent is *committed* to. (This process may recurse, so that the agent then again considers its options, at a lower level of abstraction, choosing between them, and so on, until ultimately it arrives at intentions that are directly executable.)

In the philosophy literature, the term *intentional system* is used to refer to a system whose behaviour can be predicted and explained in terms of *attitudes* such as belief, desire and intention [38]. The rationale for this approach is that, in everyday life, we use a *folk psychology* to explain and predict the behaviour of complex intelligent systems: people. For example, we use statements such as *Michael intends to write a paper* in order to explain Michael's behaviour. Once told this statement, we expect to find Michael shelving other commitments and developing a plan to write the paper; we would expect him to spend a lot of time at his computer; we would not be surprised to find him in a grumpy mood; but we *would* be surprised to find him at a late night party.

This *intentional stance*, whereby the behaviour of a complex system is understood via the attribution of attitudes such as believing and desiring, is simply an *abstraction tool*. It is a convenient shorthand for talking about complex systems, which allows us to succinctly predict and explain their behaviour without having to understand how they actually work. Now, much of computer science is concerned with looking for good abstraction mechanisms, since these allow system developers to manage complexity with greater ease: witness procedural abstraction, abstract data types and, most recently, objects. So why not use the intentional stance as an abstraction tool in computing – to explain, understand and, crucially, *program* complex computer systems?

This idea of programming computer systems in terms of 'mentalistic' notions such as belief, desire and intention is a key component of the BDI model. The concept was first articulated by Yoav Shoham, in his *agent-oriented programming* (AOP) proposal [87]. There seem to be a number of arguments in favour of AOP.

- First, it offers us a familiar, non-technical way to talk about complex systems. We need no formal training to understand mentalistic talk: it is part of our everyday linguistic ability.

- Secondly, AOP may be regarded as a kind of 'post-declarative' programming. In procedural programming, saying what a system should do involves stating precisely *how* to do it, by writing a detailed algorithm. Procedural programming is difficult because it is hard for people to think in terms of the detail required. In declarative programming (*à la* Prolog), the aim is to reduce the emphasis on control aspects: we state a goal that we want the system to achieve, and let a built-in control mechanism (e.g. goal-directed refutation theorem proving) figure out what to do in order to achieve it. However, in order to successfully write efficient or large programs in a language like Prolog, it is necessary for the programmer to have a detailed understanding of how the built-in control mechanism works. This conflicts with one of the main goals of declarative programming: to relieve the user of the need to deal with control issues. In AOP, the idea is that, as in declarative programming, we state our goals, and let the built-in control mechanism figure out what to do in order to achieve them. In this case, however, the control mechanism implements some model of rational agency. Hopefully, this computational model corresponds to our own intuitive understanding of (say) beliefs and desires, and so we need less special training to use it. Ideally, as AOP programmers, we would not be concerned with *how* the agent achieves its goals. (The reality, as ever, does not quite live up to the ideal!)

2.2   Practical Reasoning

So, the key data structures in our agents will be beliefs, desires and intentions. How does an agent with beliefs, desires and intentions go from these to its actions? The particular model of decision-making underlying the BDI model is known as *practical reasoning*. Practical reasoning is *reasoning directed towards actions* – the process of figuring out what to do:

> Practical reasoning is a matter of weighing conflicting considerations for and against competing options, where the relevant considerations are provided by what the agent desires/values/cares about and what the agent believes. [21, p. 17]

Human practical reasoning seems to consist of two distinct activities: *deliberation* (fixing upon states of affairs that we want to achieve, i.e. our intentions); and *means-ends reasoning* (deciding how to act so as to bring about our intentions).

Deliberation and Intentions

The deliberation process results in an agent adopting *intentions*. Since intentions play such a key role in the BDI model, it is worth spending some time discussing their properties.

The most obvious role of intentions is that they are *pro-attitudes* [21, p. 23]: they tend to lead to action. If you have an intention to write a book, then we would expect you to make some attempt to achieve this intention. If you sit back and do nothing, then we might be inclined to say you never had any such intention at all. Bratman notes that intentions play a much stronger role in influencing action than other pro-attitudes, such as mere desires:

> My desire to play basketball this afternoon is merely a potential influencer of my conduct this afternoon. It must vie with my other relevant desires [. . .] before it is settled what I will do. In contrast, once I intend to play basketball this afternoon, the matter is settled: I normally need not continue to weigh the pros and cons. When the afternoon arrives, I will normally just proceed to execute my intentions. [21, p. 22]

The second main property of intentions is that they *persist*. If you adopt an intention, then you should persist with this intention and attempt to achieve it. For if you immediately drop your intentions without devoting any resources to achieving them, we would be inclined to say that you never really had an intention in the first place, as we saw in an earlier example. Of course, you should not persist with your intention for too long – if it becomes clear that you will never become an academic, then it is only rational to drop your intention to do so. Similarly, if the reason for having an intention goes away, then it is rational to drop the intention. For example, if you adopted the intention to become an academic because you believed it would be an easy life, but then discover that this is not the case, then the justification for the intention is no longer present, and you should drop it. Moreover, if you initially fail to achieve an intention, then we would expect you to *try again* – we would not expect you to simply give up.

The third main property of intentions is that, once you have adopted an intention, the very fact of having this intention will constrain your future practical reasoning. While you hold some particular intention, you will not subsequently entertain options that are *inconsistent* with that intention. Thus, intentions provide a 'filter of admissibility' in that they constrain the space of possible intentions that an agent needs to consider.

Finally, intentions are closely related to beliefs about the future. In particular, intending something implies that you believe that this thing is in principle possible, and, moreover, that 'under normal circumstances', you will succeed with your intentions. However, you might also believe it is possible that your intention might *fail*.

Means-Ends Reasoning

Means-ends reasoning is the process of deciding how to achieve an end (i.e. an intention that you have chosen) using the available means (i.e. the *actions* that you can perform in your environment). Means-ends reasoning is perhaps better known in the AI community as *planning* [49]. A *planner* is a system that takes as input representations of:

1. A *goal*, or *intention*: something that the agent wants to achieve.

2. The agent's current *beliefs* about the *state of the environment*.

3. The *actions* available to the agent.

As output, a planning algorithm generates a *plan*. A plan is a course of action – a 'recipe'. If the planning algorithm does its job correctly, then if the agent executes this plan ('follows the recipe') from a state in which the world is as described in (2), then once the plan has been completely executed, the goal/intention in (1) will be achieved [65].

The original focus in AI was on the production of plans from 'first principles' [4]. By this, we mean the assembly of a complete course of action (essentially, a program) in which the atomic components are the actions available to the agent. The major problem with this approach is that it is computationally very costly: even with hopelessly restrictive assumptions, planning of this type is PSPACE-complete, and hence (under standard complexity theoretic assumptions) 'harder' than NP-hard problems such as the travelling salesman problem [49, pp. 55 – 67]. With less restrictive, more realistic assumptions, planning is computationally much worse even than this. Although significant advances have been made in the development of efficient algorithms for this type of planning, this underlying complexity makes many researchers doubt whether such planning algorithms might be used in decision-making environments where an agent must plan and act in real time.

An alternative approach is to completely reject plans as the components of decision-making, and to seek instead alternative decision-making models [24]. Although this is an interesting direction, it is somewhat tangential to our concerns in this book, and so we simply refer the reader to, e.g. [103, 104], for further discussion and relevant references.

Instead, we will focus on one simple idea that has proven to be quite powerful in practice: the idea that a programmer develops collections of partial plans for an agent offline (i.e. at design time), and the task of the agent is then to assemble these plans for execution at run time, to deal with whichever goal the agent is currently working towards. This approach may sound like a simple variation of first principles planning (instead of assembling actions, we assemble plans); and moreover, it seems to lack the flexibility of first principles planning (the ability of an agent to deal with circumstances is dependent on the plans coded for it by the agent programmer). Nevertheless, experience tells us that the approach can work well in practice, and this is the model that is implemented in AgentSpeak.

2.3 A Computational Model of BDI Practical Reasoning

Suppose we want to *implement* an agent based on the practical reasoning model we outlined above, i.e. making use of deliberation and means-ends reasoning. What is the program going to look like? As a first attempt, we might imagine the following control loop, in which the agent continually:

- looks at the world, and updates beliefs on this basis;

- deliberates to decide what intention to achieve;

- uses means-ends reasoning to find a plan to achieve this intention;

- executes the plan.

If we go back to our desirable properties of an agent from Chapter 1, however, we find this basic control loop has many problems. First, consider the issue of how *committed* such an agent is to its chosen intentions and its chosen plans. Specifically, the agent is too strongly committed, in the following sense. Once chosen, the agent remains committed to an intention (i.e. some state of affairs to achieve) until it has completely executed a plan to achieve this state of affairs, even if, while the agent is executing the plan, the intention becomes unachievable. Similarly, once the agent has selected a plan, it remains committed to it even if environmental circumstances render the plan useless.

With these issues in mind, we can, after some consideration, come up with a refined version of the agent control loop, which tries to achieve a *rational balance* between its commitments [29]: with this control loop, the agent remains committed to its plans until either they have been fully executed, the intention for which they were developed has been achieved or is believed to be unachievable, or they are believed to be no longer any use. The agent remains committed to its intentions until it believes it has achieved them, they are impossible to achieve, or there

```
1.   B ← B₀;        /* B₀ are initial beliefs */
2.   I ← I₀;        /* I₀ are initial intentions */
3.   while true do
4.      get next percept ρ via sensors;
5.      B ← brf(B, ρ);
6.      D ← options(B, I);
7.      I ← filter(B, D, I);
8.      π ← plan(B, I, Ac);  /* Ac is the set of actions */
9.      while not (empty(π) or succeeded(I, B) or impossible(I, B)) do
10.        α ←  first element of π;
11.        execute(α);
12.        π ←  tail of π;
13.        observe environment to get next percept ρ;
14.        B ← brf(B, ρ);
15.        if reconsider(I, B) then
16.           D ← options(B, I);
17.           I ← filter(B, D, I);
18.        end-if
19.        if not sound(π, I, B) then
20.           π ← plan(B, I, Ac)
21.        end-if
22.     end-while
23.  end-while
```

Figure 2.1 Overall control loop for a BDI practical reasoning agent.

is something else more fruitful that the agent could be working towards. Figure 2.1 gives the pseudo-code for the control cycle of such an agent.

In this loop, the variables B, D and I hold the agent's current beliefs, desires and intentions respectively. As described above, the agent's beliefs represent information that the agent has about its environment, while the intentions variable I contains states of affairs that the agent has chosen and committed to. Finally, D contains the agent's desires: these are the *options* that the agent is considering, and may be thought of as *candidates* to be intentions.

The basic control loop for the agent is on lines (3) – (23): roughly, this corresponds to the overall loop that we informally introduced above. However, it is somewhat more refined, as follows.

Within the basic control loop, the agent observes its environment to get the next percept. Then, in line (5), it updates its beliefs (i.e. its current information about the environment) via the *belief revision function*, which we denote by $brf(\ldots)$. This function takes the agent's current beliefs and the new percept (ρ), and returns the agent's new beliefs – those that result from updating B with ρ.

At line (6), the agent determines its desires, or options, on the basis of its current beliefs and intentions, using the function *options*($\ldots$). The agent then chooses between these options, selecting some to become intentions (line 7), and generating a plan to achieve the intentions, via its *plan*($\ldots$) function (line 8).

The inner loop on lines (9) – (22) captures the execution of a plan to achieve the agent's intentions. If all goes well, then the agent simply picks off each action in turn from its plan and executes it, until the plan π is *empty*($\ldots$) (i.e. all the actions in the plan have been executed). However, after executing an action from the plan – line (11) – the agent pauses to observe its environment, again updating its beliefs. It then asks itself whether, in principle, it is worth *reconsidering* its intentions (i.e. spending time deliberating over them again): it makes this decision via the *reconsider*($\ldots$) function – lines (15) – (18). Roughly, we usually want to reconsider our intentions *only* when such reconsideration would actually lead to a change of intentions. If reconsideration does not lead to a change in intentions, then usually, we would be better off not wasting time and computational effort in deliberation: we should simply get on with our intentions – see [59, 84, 85].

Finally, irrespective of whether or not it decided to deliberate, it asks itself whether or not the plan it currently has is *sound* with respect to its intentions (what it wants to achieve with its plan) and its beliefs (what it thinks the state of the environment is) – line (19). If it believes the plan is no longer a sound one, then it replans – line (20).

2.4   The Procedural Reasoning System

The control loop in Figure 2.1 is still quite a long way from an actual implementation: we have not explained how the various functions might be implemented, or what the contents of B, D and I actually are. The Procedural Reasoning System (PRS), originally developed at Stanford Research Institute by Michael Georgeff and Amy Lansky, was perhaps the first agent architecture to explicitly embody the belief – desire – intention paradigm, and has proved to be one of the most durable approaches to developing agents to date. The PRS has been re-implemented several times since the mid-1980s, for example in the Australian AI Institute's dMARS system [40], the University of Michigan's C++ implementation UM-PRS, and a Java version called JAM! [54]. JACK is a commercially available programming language, which extends the Java language with a number of BDI features [25]. An illustration of the PRS architecture is shown in Figure 2.2.

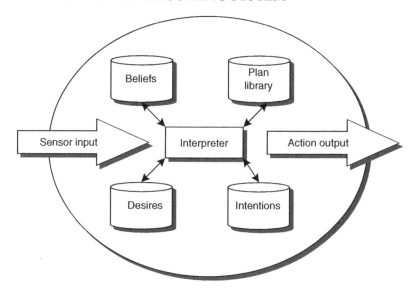

Figure 2.2 The Procedural Reasoning System (PRS).

In the PRS, an agent does no planning from first principles. Instead, it is equipped with a library of pre-compiled plans. These plans are manually constructed, in advance, by the agent programmer. Plans in the PRS each have the following components:

- a *goal* – the post-condition of the plan;

- a *context* – the pre-condition of the plan; and

- a *body* – the 'recipe' part of the plan – the course of action to carry out.

The goal of a PRS plan defines 'what the plan is good for', i.e. the things that it achieves. In conventional computer science terms, we can think of the goal as being the *post-condition* of the plan.

The *context* part of a PRS plan defines the *pre-condition* of the plan. Thus, the context of a plan defines what must be true of the environment in order for the plan to be successful.

The body of a PRS plan, however, is slightly unusual. We have so far described plans as courses of action, and in the BDI agent control loop above, we assumed that a plan was simply a list of actions. Executing such a plan simply involves executing each action in turn. Such plans are possible in the PRS, but much richer kinds of plans are also possible. The first main difference is that, as well has having individual primitive actions as the basic components of plans, it is possible to have *goals*. The idea is that, when a plan includes a goal at a particular point, this means that this goal must then be achieved at this point before the remainder of the plan can be executed. It is also possible to have disjunctions of goals ('achieve ϕ or achieve ψ'), and loops ('keep achieving ϕ until ψ'), and so on.

At start-up time a PRS agent will have a collection of such plans, and some initial beliefs about the world. Beliefs in the PRS are represented as Prolog-like facts – essentially, as atomic formulæ of first-order logic. In addition, at start-up, the agent will typically have a top-level goal. This goal acts in a rather similar way to the 'main' method in Java or C.

When the agent starts up, the goal to be achieved is pushed onto a stack, called the *intention stack*. This stack contains all the goals that are pending achievement. The agent then searches through its plan library to see what plans have the goal on the top of the intention stack as their post-condition. Of these, only some will have their pre-condition satisfied, according to the agent's current beliefs. The set of plans that (i) achieve the goal, and (ii) have their pre-condition satisfied, become the possible *options* for the agent.

The process of selecting between different possible plans is, of course, deliberation, a process that we have already discussed above. There are several ways of deliberating between competing options in PRS-like architectures. In the original PRS, deliberation is achieved by the use of *meta-level plans*. These are literally plans about plans. They are able to modify an agent's intention structures at run-time, in order to change the focus of the agent's practical reasoning. However, a simpler method is to use *utilities* for plans. These are simple numerical values; the agent simply chooses the plan that has the highest utility.

The chosen plan is then executed in its turn; this may involve pushing further goals onto the intention stack, which may then in turn involve finding more plans to achieve these goals, and so on. The process bottoms-out with individual actions that may be directly computed (e.g. simple numerical calculations). If a particular plan to achieve a goal fails, then the agent is able to select another plan to achieve this goal from the set of all candidate plans.

The AgentSpeak language, introduced by Rao [80], represents an attempt to distill the key features of the PRS into a simple, unified programming language. Rao's motivation was that the PRS system (and its successors), while being practically useful systems for the implementation of large scale systems, were some distance from the various theoretical models of BDI systems that were being proposed at the time. He wanted a programming language that provided the key features of PRS, but in a sufficiently simple, uniform language framework that it would be possible to investigate it from a theoretical point of view, for example by giving it a formal semantics.

2.5 Agent Communication

The PRS model, and the AgentSpeak language in turn, are primarily concerned with the internal structure of decision making, and in particular, the interplay between the creation of (sub-)goals, and the execution of plans to achieve these (sub-)goals. The twin issues of *communication* and *multi-agent interaction* are not addressed within the basic architecture. This raises the question of how such issues might be dealt with within the architecture. As BDI theory is based on the philosophical literature on practical reasoning [20], agent communication in multi-agent systems is typically based on the speech-act theory, in particular the work of Austin [8] and Searle [86].

Speech-act theory starts from the principle that language is action: a rational agent makes an utterance in an attempt to change the state of the world, in the same way that an agent performs 'physical' actions to change the state of the world. What distinguishes speech acts from other ('non-speech') actions is that the domain of a speech act – the part of the world that the agent wishes to modify through the performance of the act – is typically limited the mental state(s) of the hearer(s) of the utterance. Thus examples of speech acts might be to change your beliefs, desires or intentions.

Speech acts are generally classified according to their *illocutionary force* – the 'type' of the utterance. In natural language, illocutionary forces are associated with utterances (or locutionary acts). The utterance 'the door is open', for example, is generally an 'inform' or 'tell' action. The *perlocutionary force* represents what the speaker of the utterance is attempting to achieve by performing the act. In making a statement such as 'open the door', the perlocutionary force will generally be the state of affairs that the speaker hopes to bring about by making the utterance; of course, the *actual* effect of an utterance will be beyond the control of the speaker. Whether I choose to believe you when you inform me that the door is open depends upon how I am disposed towards you. In natural language, the illocutionary force and perlocutionary force will be implicit within the speech act and its context. When the theory is adapted to agent communication, however, the illocutionary forces are made explicit, to simplify processing of the communication act. The various types of speech acts are generally referred to as 'performatives' in the context of agent communication.

Searle identified various different types of speech act:

- *representatives* – such as *informing*, e.g. 'It is raining';

- *directives* – attempts to get the hearer to do something, e.g. 'Please make the tea';

- *commisives* – which commit the speaker to doing something, e.g. 'I promise to . . .';

- *expressives* – whereby a speaker expresses a mental state, e.g. 'Thank you!';

- *declarations* – such as declaring war or christening.

There is some debate about whether this typology of speech acts is ideal. In general, a speech act can be seen to have two components:

- a *performative verb*, e.g. request, inform, etc;

- *propositional content*, e.g. 'the door is closed'.

With the same content but different performatives, we get different speech acts:

- performative = request
  content = 'the door is closed'
  speech act = 'please close the door'

- performative = inform
  content = 'the door is closed'
  speech act = 'the door is closed!'

- performative = inquire
  content = 'the door is closed'
  speech act = 'is the door closed?'

Notice that the *content* of a speech act is not just a list of arguments, in the way that a Java method simply has a list of parameters. Rather, the content is a *proposition*: a statement that is either true or false. In this way, speech act communication is *knowledge-level* communication, and is very different to communication via method invocation. As we see below, *agent communication languages* take their inspiration from this model of speech acts.

Agent Communication Languages: KQML and FIPA

The Knowledge Query and Manipulation Language (KQML), developed in the context of the 'Knowledge Sharing Effort' project [44], was the first attempt to define a practical agent communication language that included high-level (speech-act based) communication as considered in the distributed artificial intelligence literature. KQML is essentially a knowledge-level messaging language [62, 69]. It defines a number of performatives, which make explicit an agent's intentions in sending a message. For example, the KQML performative `tell` is used with the intention of changing the receiver's *beliefs*, whereas `achieve` is used with the intention of changing the receiver's *goals*. Thus the performative label of a KQML message explicitly identifies the intent of the message sender.

Here is an example KQML message:

```
(ask-one
    :content  (PRICE IBM ?price)
    :receiver stock-server
    :language LProlog
    :ontology NYSE-TICKS
)
```

The intuitive interpretation of this message is that the sender is asking about the price of IBM stock. The performative is `ask-one`, which an agent will use to ask another agent a question where exactly one reply is needed. The various other components of this message represent its attributes. The most important of these is the `:content` field, which specifies the message content. In this case, the content simply asks for the price of IBM shares. The `:receiver` attribute specifies the intended recipient of the message, the `:language` attribute specifies that the language in which the content is expressed is called `LProlog` (the recipient is assumed to 'understand' `LProlog`), and the final `:ontology` attribute defines the *terminology* used in the message. The idea is that an ontology is an agreed specification of all the terms that might be used in a message. Such a commonly agreed specification is needed so that all the agents involved in a conversation know that, when they use a piece of terminology in a particular way, the other agents are also using it in the same way. Ontologies are an important topic in computer science at the time of writing, although they are somewhat tangential to the issues of this book. See [7] for an introduction to common frameworks for defining ontologies.

The FIPA standard for agent communication[1] was released in 2002. The goal of the FIPA organisation was to develop a coherent collection of standards relating to agent communication, management and use, and the agent communication standard was just one component of the FIPA suite. This standard is in fact closely based on KQML, differing in its performative set and semantics. The main goal was to simplify and rationalise the performative set as much as possible, and to address the issue of *semantics*, a somewhat problematic issue for agent communication languages.

Perhaps the first serious attempt to define the semantics of KQML was made by Labour and Finin [61]. This work built on the pioneering work of Cohen and Perrault, on an action-theoretic semantics of natural language speech acts [31]. The key insight in Cohen and Perrault's work was that, if we take seriously the idea of utterances as action, then we should be able to apply a formalism for reasoning about action to reasoning about utterances. They used a STRIPS-style pre- and

---

[1] http://www.fipa.org/specs/fipa00037/SC00037J.html.

post-condition formalism to define the semantics of 'inform' and 'request' speech acts (perhaps the canonical examples of speech acts), where these pre- and post-conditions were framed in terms of the beliefs, desires and abilities of conversation participants. When applied by Labrou and Finin to the KQML language [61], the pre- and post-conditions defined the mental states of the sender and receiver of a KQML message before and after sending a message. For the description of mental states, most of the work in the area is based on the Cohen and Levesque's theory of intention [29, 30]. Agent states are described through mental attitudes such as belief (*bel*), knowledge (*know*), desire (*want*) and intention (*intend*). These

- Pre-conditions on the states of S and R:

    - $Pre(S)$: $bel(S,X) \wedge know(S, want(R, know(R, bel(S,X))))$
    - $Pre(R)$: $intend(R, know(R, bel(S,X)))$

- Post-conditions on S and R:

    - $Pos(S)$: $know(S, know(R, bel(S,X)))$
    - $Pos(R)$: $know(R, bel(S,X))$

- Action completion:

    - $know(R, bel(S,X))$

Figure 2.3 Semantics for *tell* [61].

- Pre-conditions on the states of S and R:

    - $Pre(S)$: $want(S, know(S,Y)) \wedge know(S, intend(R, process(R,M)))$ where Y is either $bel(R,X)$ or $\neg bel(R,X)$ and M is $ask\_if(S,R,X)$
    - $Pre(R)$: $intend(R, process(R,M))$

- Post-conditions about R and S:

    - $Pos(S)$: $intend(S, know(S,Y))$
    - $Pos(R)$: $know(R, want(S, know(S,Y)))$

- Action completion:

    - $know(S,Y)$

Figure 2.4 Semantics for *ask-if* [61].

mental attitudes normally have propositions (i.e. symbolic representations of the state of the world) as arguments. Figures 2.3 and 2.4 give a semantics for the KQML performatives $tell(S, R, X)$ (S tells R that S believes that X is true) and $ask\_if(S, R, X)$ (S asks R if R believes that X is true), in the style of Labrou and Finin [61].

3

The *Jason* Agent Programming Language

We usually find it convenient to make a distinction between an *agent program* and an *agent architecture*. The agent architecture is the software framework within which an agent program runs. The PRS is an example of an agent architecture; the plans are the program that inhabits this architecture. We write the program that will direct the agent behaviour, but much of what the agent effectively does is determined by the architecture itself, without the programmer having to worry about it. The language interpreted by *Jason* is an extension of AgentSpeak, which is based on the BDI architecture. Thus, one of the components of the agent architecture is a belief base, and an example of one of the things that the interpreter will be doing constantly, without it being explicitly programmed – even though this can be customised, which we shall see in later chapters – is to perceive the environment and update the belief base accordingly. Another important component is the agent's *goals*, which are achieved by the execution of *plans*.

Recall that practical BDI agents are *reactive planning systems*. These are reactive systems in the sense described previously: systems that are not meant to compute the value of a function and terminate, but rather designed to be permanently running, reacting to some form of 'events'. The way reactive planning systems react to such events is by executing plans; plans are courses of action that agents commit to execute so as to handle such events. The actions in turn change the agent's environment, in such a way that we can expect the agent's goals to be achieved. Again, there is a component of the agent architecture which will be responsible for effecting the changes in the environment based on the *choices* on the course of actions made by the interpretation of the agent program.

The interpretation of the agent program effectively determines the agent's *reasoning cycle*. An agent is constantly perceiving the environment, reasoning about how to act so as to achieve its goals, then acting so as to change the environment. The (practical) reasoning part of the agent's cyclic behaviour, in an AgentSpeak agent, is done according to the plans that the agent has in its *plan library*. Initially, this library is formed by the plans the programmer writes as an AgentSpeak program.

We shall see in detail all the components of the agent architecture and how the interpreter works in the next chapter. In this chapter, we describe all the language constructs that programmers can use, which can be separated into three main categories: beliefs, goals and plans. There are various different types of constructs used in each category; we will first introduce all of them and then give an example using as many language constructs as possible.

3.1 Beliefs

The very first thing to learn about AgentSpeak agents is how to represent the agents' beliefs. As we mentioned above, an agent has a *belief base*, which in its simplest form is a collection of literals, as in traditional logic programming. In programming languages inspired by logics, information is represented in symbolic form by *predicates* such as:

```
tall(john).
```

which expresses a particular *property* – that of being *tall* – of an object or individual, in this case '*John*' (the individual to whom the term `john` refers). To represent the fact that a certain relationship holds between two or more objects, we can use a *predicate* such as:

```
likes(john, music).
```

which, pretty obviously, states that John likes music. A *literal* is such a predicate or its negation – we say more about negation below.

However, unlike in classical logic, we are here referring to the concept of *modalities of truth* as in the modal logic literature, rather than things that are stated to be true in absolute terms. When a formula such as `likes(john, music)` appears in an agent's belief base, that is only meant to express the fact that the agent currently *believes* that to be true; it might well be the case that John does not like music at all.

We shall not dwell too long on the issue of symbolic representation of logical statements; the vast literature on logic and logic programming [19, 28, 57, 63, 66, 92] is the best place to find more details about this. However, we introduce below some of the basic concepts, just so that the reader unfamiliar with a language such

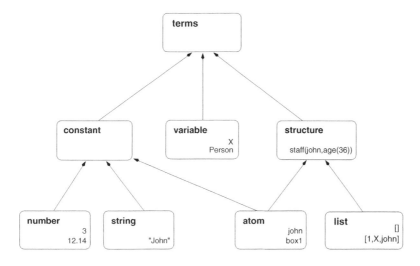

Figure 3.1 Types of AgentSpeak terms in *Jason*.

as Prolog can carry on reading this book. We concentrate here on presenting the main language constructs in logic programming, and Chapter 4 will give some examples which can help with understanding other things such as the use of logical variables and the important issue of *unification*. A summary of all the notions used to define the basic terms of the logic-based language used in this book is given in Figure 3.1.

Basics of logic programming

As in Prolog, any symbol (i.e. a sequence of characters) starting with a lower-case letter is called an *atom*, which is used to represent particular individuals or objects – they are the equivalent of 'constants' in first-order logic. For example, john is an atom (which can be used to represent a real person called John). A symbol starting with an uppercase letter is interpreted as a *logical variable*. So, while john is an atom, Person is a variable. Besides atoms, numbers and strings (which are represented as usual in programming languages) are also classified as *constants*.

Initially variables are *free* or *uninstantiated* and once *instantiated* or *bound* to a particular value, they maintain that value throughout their *scope* – in the case of AgentSpeak, we shall see that the scope is the *plan* in which they appear. Variables are bound to values by *unification*; a formula is called *ground* when it has no more uninstantiated variables. Using the examples above, a variable Person could be bound to the atom john. More generally, variables can be bound to a constant (i.e. an atom, a number, or a string), or to a more complex data type called *structure*. The word *term* is used to refer to a constant, a variable or a structure.

A *structure* is used to represent complex data, for example `staff("Jomi Fred Hübner", 145236, lecturer, married, wife(ilze), kids([morgana,thales]), salary(133987.56))` could be used to represent the information about a member of staff in a university. Structures start with an atom (called the *functor*) and are followed by a number of terms (called *arguments*) separated by commas and enclosed in parentheses. Observing the predicates (called *facts* in Prolog) shown earlier to exemplify an agent's beliefs, we can see that they have exactly the same format as structures. The difference is purely semantical: a structure is used as term, to represent an individual or object, whereas a fact is used to represent a logical proposition – i.e. a formula that is to be interpreted as being either true or false. The number of arguments of a predicate/structure is important, and is called its *arity*. A particular structure is referred to by its functor and its arity: the structure above, for example, is referred to by '`staff/7`', as it must have always exactly seven different terms as arguments (otherwise it is considered a different structure altogether).

A special type of structure was used in the example above: `[morgana,thales]`. Although internally this is represented as a structure, this special notation, called a *list*, is very useful in logic programming. There are special operations for lists; in particular '`|`' can be used to separate the first item in a list from the list of all remaining items in it. So, if we match (more precisely, unify) a list such as `[1,2,3]` with list `[H|T]`, `H` (the head) is instantiated with `1` and `T` (the tail) with the list `[2,3]`. The Prolog syntax for lists is also used in *Jason*.

Figure 3.1 shows the classification of the various types of terms available in the language interpreted by *Jason*. Terms are then used to compose more complex types of formulæ, as shown later in Figure 3.2.

Annotations

Perhaps one important difference in the syntactical representation of logical formulæ in *Jason* compared with traditional logic programming is the use of *annotations*. These are complex terms (cf. Prolog structures) providing details that are strongly associated with one particular belief. Annotations are enclosed in square brackets immediately following a literal, for example:

```
busy(john)[expires(autumn)].
```

which, presumably, means that the agent believes that John is busy, but as soon as autumn starts, that should be no longer believed to hold. Note that the annotation `expires(autumn)` is giving us further information about the `busy(john)` belief in particular. Readers familiar with logic will see that annotations of this form do not increase the expressive power of the language. We could equally well have written something like this:

```
busy_until(john,autumn).
```

to mean pretty much the same thing. Or if we wanted to have the possibility of marking down when each single belief is to be expired (and removed from the belief base), we could have used `annot(b,expires(...))` for each belief `b` that has a date to expire, where `annot` would be a special predicate symbol to be used by customisations of the *Jason* interpreter. However, there are two advantages of the way annotations are used in *Jason*. The first is simply the elegance of the notation. Elegance might be irrelevant logically, but for many programmers it is of great importance! The fact that the details about a particular belief are organised in such a way so as to be visually linked to belief itself makes the belief base considerably more readable. The second advantage is that this representation also facilitates management of the belief base. For example, the `expires` annotation is presumably meant to influence the duration of time that the `busy(john)` belief will be kept in the belief base.

Note that the `expires` annotation means absolutely nothing to the *Jason* interpreter. One should not expect *Jason* to delete that belief simply because the programmer annotated when the belief expires. On the other hand, it would be very easy for the programmer to customise, by doing a little (and very simple) Java programming, an agent's belief base (or indeed the agent's belief update function) so that the `busy(john)[expires(autumn)]` belief was automatically removed as soon as the agent believed `autumn` to be true. We shall see how this can be done in Chapter 7.

There are other belief annotations, however, which do have specific meanings for the interpreter, and very important ones, for that matter. In particular, the `source` annotation is used to record what was the source of the information leading to a particular belief. In fact, the `source` annotation is the reason why annotations were created in the first place, even though they later turned out to be useful for so many (application-specific) programming tasks.

There are three different types of information source for agents in multi-agent systems:

> **perceptual information:** an agent acquires certain beliefs as a consequence of the (repeated) sensing of its environment – a symbolic representation of a property of the environment as perceived by an agent is called a *percept*;

communication: as agents communicate with other agents in a multi-agent system, they need to be able to represent the information they acquired from other agents, and it is often useful to know exactly which agent provided each piece of information;

mental notes: it makes certain programming tasks easier if agents are able to remind themselves of things that happened in the past, or things the agent has done or promised: beliefs of this type are added to the belief base by the agent itself as part of an executing plan, and are used to facilitate the execution of plans that might be selected at a later stage. The best way to think of this type of belief is as a 'mental note'; that is, things the agent itself will need to remember in certain future circumstances.

One of the things that is perhaps different in *Jason* compared with other agent programming languages is the way *Jason* makes the source of information explicit in the belief base. For beliefs originating from perceptual information, the interpreter automatically adds an annotation `source(percept)`, which tells us that this particular belief was perceived by the agent itself rather than, for example, communicated. An annotation `source(self)` is used for mental notes (these are explained in more detail on page 52). Any other source of information annotated by the interpreter will be the name of the agent which sent a communication message that provided the information leading to that belief. For example, `likes(john,music)[source(john)]` says that our agent believes `john` likes music because agent `john` itself informed our agent of this (which implies that, at least at the time that belief was acquired, there was an agent named `john` which was a member of the same *society* as our agent).

Strong Negation

Negation is the source of many difficulties in logic programming languages. A popular approach to handling negation in logic programming is to use the *closed world assumption* and *negation as failure*. The closed world assumption may be understood as follows:

Anything that is neither known to be true, nor derivable from the known facts using the rules in the program, is assumed to be false.

The only type of negation in this context, given by the 'not' operator, is such that the negation of a formula is true if the interpreter *fails* to derive the formula using the facts and rules in the program (rules are also available in our language, as we see later). Another type of negation, denoted here by the '~' operator, and called *strong negation*, is used to express that an agent *explicitly believes something to be false.* For example, the belief `colour(box1,white)` means that the agent believes the colour of `box1` *is* white, whereas the belief `~colour(box1,white)` means that the agent believes that it is *not* the case that the colour of `box1` is white.

vanced

Under the closed world assumption, anything not explicitly believed to be true by the agent is believed to be false. However, this assumption is too simple and too strong for many applications. For open systems (such as, e.g., the Web), it is significantly easier to produce a model for an agent's beliefs if one is able to refer to the things the agent believes to be true, the things the agent explicitly believes to be false, as well as the things the agent is ignorant about. For this, it helps to have a different type of negation, which here is denoted by '~'. So, if `not p` and `not ~p` are true (meaning that neither p nor ~p are in – or can be derived from – the agent's belief base) we can conclude that the agent has no information about whether p is true or not. We postpone giving a more concrete example for this until we have seen how we write plans in AgentSpeak.

Recall that some of the beliefs in an agent's belief base will have been placed there by the agent repeatedly sensing the environment. Therefore, if the closed world assumption is to be used, make sure that the program does not use strong negation and also make sure that the percepts the agent obtains when sensing the environment (i.e. by its perception function) only contain *positive literals*; that is, strong negation cannot be used to model the environment either. Chapter 5 will discuss simulated environments, and Section 7.3 will discuss the customisation of the perception function.

Now, here is an example which includes all we have learnt about the belief base so far. Consider the following excerpt of `maria`'s belief base:

```
colour(box1,blue)[source(bob)].
~colour(box1,white)[source(john)].
colour(box1,red)[source(percept)].
colourblind(bob)[source(self),degOfCert(0.7)].
lier(bob)[source(self),degOfCert(0.2)].
```

The example tells us that agent `maria`, whose belief base is as above, was informed by agent `bob` that a particular box named `box1` is blue (this is what the first line of the example above says) and was informed by agent `john` that the box is *not* white (second line of the example). However, agent `maria` (later) perceived – e.g. by means of a camera – that `box1` is red, according to the third line of the example. Now, normally, and in particular in open systems where agents cannot be assumed to be trustworthy, perceptual information is more reliable than communicated information, so in principle this agent can conclude that either bob is colourblind, and therefore was genuinely mistaken about the colour of box1, or else that bob intentionally tried to deceive maria. Comparing the degrees of certainty (annotation degOfCert) that have been associated with the last two lines of the example, we can imagine that maria has had some form of interaction with bob and is fairly confident of its trustworthiness, given that maria seems to be much more certain of the possibility that bob is colourblind than of it being a liar. Note the source(self) annotations in the last two lines, which tell us that those beliefs were added to the belief base by the agent itself (how an agent can add such derived beliefs to the belief base will be seen below on page 52). Again, we should emphasise that degOfCert means nothing to the interpreter, that is to say that any operations on, or considerations of, degrees of certainty need to taken care of by the programmer rather than to expect the interpreter to do this automatically.

Advanc

Nested annotations

It is also possible to use *nested annotations*. For example, in bob's belief loves(maria,bob)[source(john)[source(maria)]] there is an annotation (source(maria)) to source(john) which is itself an annotation. This would happen, for example, if agent john sent a message to bob with content loves(maria,bob)[source(maria)]. What this represents is that agent john is telling bob that maria told him that she loves bob. Note further that, if the multi-agent system assumes the principle of *veracity* (see [103]), meaning that the designer can assume that all agents in the system only inform other agents of things they themselves believe to be true, then loves(maria,bob)[source(john)[source(maria)]] could also represent a nested belief modality (for the reader interested in modal logics): after this communication, bob believes that john believes that maria believes that she loves bob.

Rules

Readers with experience of Prolog will be familiar with the idea of rules, which allow us to conclude new things based on things we already know. Including such rules in an agent's belief base can simplify certain tasks, for example in making certain conditions used in plans (as discussed in Section 3.3 below) more succinct. However, for simple applications or for those who are not Prolog experts, it may be preferable not to use rules altogether. If they are used, Prologers will note that the notation is slightly different, the reason being that the types of formulæ that can appear in the body of the rules also appear elsewhere in the AgentSpeak language (in a plan 'context'), so we use that same notation here for the sake of homogeneity.

Consider the following rules:

```
likely_colour(C,B)
    :- colour(C,B)[source(S)] & (S == self | S == percept).

likely_colour(C,B)
    :- colour(C,B)[degOfCert(D1)] &
       not (colour(_,B)[degOfCert(D2)] & D2 > D1) &
       not ~colour(C,B).
```

The first rule says that the most likely colour of a box is either that which the agent deduced earlier, or the one it has perceived. If this fails, then the likely colour should be the one with the highest degree of certainty associated with it, provided there is no strong evidence the colour of the box is *not* that one. To the left of the ':-' operator, there can be only one literal, which is the conclusion to be made if the condition to the right is satisfied (according to the agent's current beliefs).

The type of reasoning that the agent is doing in the example above is referred to in the literature as *theoretical reasoning*; that is, reasoning over an abstract representation of things known in order to derive further knowledge. We mentioned in the initial chapters that the underlying model of decision-making in BDI agents is called *practical reasoning* (i.e. reasoning directed towards actions); in AgentSpeak, this is captured by the agent's *plans*, as defined in Section 3.3.

Although AgentSpeak programmers will manipulate and inspect the belief base in various ways, in AgentSpeak, all that has to be provided are the *initial beliefs* (and rules, if any). That is, the source code tells the interpreter which beliefs should be in the agent's belief base when the agent first starts running, even before the agent senses the environment for the very first time. Unless specified otherwise, all beliefs that appear in an agent's source code are assumed to be 'mental notes'. Thus, if we have likes(john, music) in the source code, when inspecting the agent's belief base we will note that the belief is in fact represented as likes(john, music)[source(self)]; we mention in Section 4.3 how to 'inspect' an agent's mental state in *Jason*. If by any chance the agent has to have an initial belief that is to be treated as if it had been acquired, e.g. by sensing the environment, that needs to be made explicit, for example as in tall(john)[source(percept)]. We shall see in Sections 4.1 and 7.2 how different types of beliefs are treated differently when belief update/revision takes place.

We now turn from belief to another key notion in AgentSpeak programming: that of a *goal*.

3.2   Goals

In agent programming, the notion of *goal* is fundamental. Indeed, many think this is the essential, defining characteristic of agents as a programming paradigm. Whereas beliefs, in particular those of a perceptual source, express properties that are believed to be true of the world in which the agent is *situated*, goals express the properties of the states of the world that the agent *wishes to bring about*. Although this is not enforced by the AgentSpeak language, normally, when representing a goal *g* in an agent program, this means that the agent is committed to act so as to change the world to a state in which the agent will, by sensing its environment, *believe* that *g* is indeed true. This particular use of goals is referred to in the agent programming literature as a *declarative goal*.

In AgentSpeak, there are two types of goals: *achievement goals* and *test goals*. The type of goal described above is (a particular use of) an achievement goal, which is denoted by the '!' operator. So, for example, we can say `!own(house)` to mean that the agent has the goal of achieving a certain state of affairs in which the agent will believe it owns a house, which probably implies that the agent does not currently believe that '`own(house)`' is true. As we shall see in the next section, the fact that an agent may adopt a new goal leads to the execution of *plans*, which are essentially courses of actions the agent expects will bring about the achievement of that goal.

Perhaps an important distinction to make is with the notion of 'goal' as used in Prolog, which is rather different from the notion of goal used above. A goal or 'goal clause' in Prolog is a conjunction of literals that we want the Prolog interpreter to check whether they can be concluded from the knowledge represented in the program; the Prolog interpreter is essentially proving that the clause is a logical consequence of the program. This is also something that we will need doing as part of our agents – that is, checking whether the agent believes a literal or conjunction of literals – and this is why there is another type of goal in AgentSpeak, called a test goal and denoted by the '?' operator. Normally, test goals are used simply to retrieve information that is available in the agent's belief base. Therefore, when we write `?bank_balance(BB)`, that is typically because we want the logical variable BB to be instantiated with the specific amount of money the agent currently *believes* its bank balance is.

More specifically, the agent could have a belief `bank_balance(100)`, in which case it is trivial to find a unification for the test goal (i.e. where BB ↦ 100), as one would expect in Prolog. Also as in Prolog, we may need to use rules in the belief base to *infer* an answer to our test goal. Furthermore, in the next section we shall see that, as for achievement goals, test goals can also lead to the execution of plans in certain circumstances.

Procedural and declarative goals

Achievement goals can be used *procedurally* or *declaratively*. In a *procedural goal*, the goal name is similar to the name of a method/procedure in traditional programming languages, which of course can be quite useful (e.g. to group together actions that are often used in the program). However, what makes the use of goals special in agent programming is when we introduce the notion of a *declarative goal*, which we already briefly mentioned above. The importance of this notion is that goals are then seen as a symbolic representation of a state of affairs that the agent has to achieve, and if an agent actually has such an explicit representation, it can then check whether having executed a piece of code (i.e. a plan) has effected, in regards to the changes in the environment, the expected result. Because multi-agent systems are particularly suitable for complex, dynamic environments, being able to make such checks is absolutely essential. Further, having such a representation means that an agent can be more naturally programmed to actively persist in attempting to achieve its goals, until the agent believes the goal has been effectively achieved. We shall return to the issue of declarative goals, and various forms of commitments that agents can have towards such goals, in Chapter 8.

In the same way that, in the actual source code, we only provide the beliefs that we wish the agent to have at the moment it first starts running, we can also provide goals that the agent will attempt to achieve from the start, if any such goals are necessary for a particular agent. That is, we can provide *initial goals* to our agent. For example, this is an agent that would start running with the long term goal of getting the house clean.

```
!clean(house). ...
```

Beliefs and goals are the two important mental attitudes that we can express in an agent source code. The third essential construct of an agent program, in the BDI style, are *plans* – the agent's know-how – and the way plans are executed is also important. It is precisely the *changes* in the agent's *beliefs* and *goals* that trigger the execution of plans. So next we see in detail how plans are written.

3.3 Plans

An AgentSpeak plan has three distinct parts: the *triggering event*, the *context*, and the *body*. Together, the triggering event and the context are called the *head* of the plan. The three plan parts are syntactically separated by ':' and '<-' as follows:

```
triggering_event : context <- body.
```

We shall explain in detail each of these parts in turn, but first we briefly describe what is the idea behind each part of a plan:

> **Triggering event.** As we saw in the initial chapters, there are two important aspects of agent behaviour: reactiveness and pro-activeness. Agents have goals which they try to achieve in the long term, determining the agent's pro-active behaviour. However, while acting so as to achieve their goals, agents need to be attentive to changes in their environment, because those changes can determine whether agents will be effective in achieving their goals and indeed how efficient they will be in doing so. More generally, changes in the environment can also mean that there are new opportunities for the agent to do things, and perhaps opportunities for considering adopting new goals that they previously did not have or indeed dropping existing goals.
>
> There are, accordingly, two types of changes is an agent's mental attitudes which are important in an agent program: changes in beliefs (which, recall, can refer to the information agents have about their environment or other agents) and changes in the agent's goals. Changes in both types of attitudes create, within the agent's architecture, the *events* upon which agents will act. Further, such changes can be of two types: *addition* and *deletion*. As we mentioned in the beginning of this chapter, plans are courses of actions that agents commit to execute as a consequence of such changes (i.e. events).
>
> The triggering event part of a plan exists precisely to tell the agent, for each of the plans in their plan library, which are the specific events for which the plan is to be used. If an event that took place matches − and we describe in detail how that matching happens in the next chapter − the triggering event of a plan, that plan might start to execute, provided some conditions are satisfied, as we see next. If the triggering event of a plan matches a particular event, we say that the plan is *relevant* for that particular event.
>
> **Context.** As for triggering events, the *context* of a plan also relates to an important aspect of reactive planning systems. We have seen that agents have goals, and plans are used to achieve them, but also that agents have to be attentive to changes in the environment. Dynamic environments are complicated to deal with first because changes in the

environment may mean we have to act further, but also because, as the environment changes, the plans that are more likely to succeed in achieving a particular goal also change.

This is why reactive planning systems postpone committing to courses of action (i.e. a plan) so as to achieve a particular goal until as late as possible; that is, the choice of plan for one of the many goals an agent has is only made when the agent is about to start acting upon it. Typically, an agent will have various different plans to achieve the same goal. (In addition, various plans for different goals can be competing for the agent's attention, but we leave this issue for the next chapter.)

The context of a plan is used precisely for checking the current situation so as to determine whether a particular plan, among the various alternative ones, is likely to succeed in handling the event (e.g. achieving a goal), given the latest information the agent has about its environment. Therefore, a plan is only chosen for execution if its context is a *logical consequence* – and we discuss later exactly what that means – of the agent's beliefs. A plan that has a context which evaluates as true given the agent's current beliefs is said to be *applicable* at that moment in time, and is a candidate for execution.

Body. The body of a plan, in general terms, is the easiest part of a plan to understand. This is simply a sequence of formulæ determining a course of action – one that will, hopefully, succeed in handling the event that triggered the plan. However, each formula in the body is not necessarily a 'straight' action to be performed by the agent's effectors (so as to change the environment). Another important construct appearing in plan bodies is that of a *goal* this allows us to say what are the (sub)goals that the agent should adopt and that need to be achieved in order for that plan to handle an event successfully. We can refer to the term *subgoals*, given that the plan where they appear can itself be a plan to achieve a particular goal – recall that the triggering events allow us to write plans to be executed when the agent has a new goal to achieve. In fact, there are other things that can appear in the body of a plan, but we see this in detail below.

Next, we introduce how exactly each part of the plan is written in an agent program.

Triggering Events

As we saw earlier, there are two types of mental attitudes: beliefs and goals. Further, there are two different types of goals: 'test' and 'achievement'. Events represent

Table 3.1   Types of triggering events.

Notation	Name
+*l*	Belief addition
−*l*	Belief deletion
+!*l*	Achievement-goal addition
−!*l*	Achievement-goal deletion
+?*l*	Test-goal addition
−?*l*	Test-goal deletion

Nomenclature for the six types of triggering events that plans can have.

changes in beliefs and goals, and the changes can be of two types: 'addition' or 'deletion'. These allows us to express six different types of triggering events for a given literal *l*, with notation and name as shown in Table 3.1.

Events for belief additions and deletions happen, for example, when the agent updates its beliefs according to its perception of the environment obtained (normally) at every reasoning cycle. Recall that the reasoning cycle and the handling of events will be discussed in detail in the next chapter.

Events due to the agent having new goals (i.e. goal additions) happen mostly as a consequence of the execution of other plans, but also as a consequence of agent communication, as we shall see in Chapter 6. When we discuss the plan body below, we shall see that executing a plan may require the agent having new goals to achieve/test. The goal deletion types of events are used for handling plan failure; we defer discussing this until Section 4.2, but keep in mind that these six types of triggering events exist.

Figure 3.2 shows how simpler formulæ are used to form the various types of AgentSpeak formulæ that have been described above (refer back to Figure 3.1 for a diagram with the various types of terms used to build formulæ).

To make the use of triggering events clearer, consider the following examples. Suppose we want to develop an agent that may need to have the goal of serving dinner, which implies that the goal will have been achieved when the agent believes that dinner is served, represented as `served(dinner)`. We will need a plan that the agent will start executing when it comes to have the goal of achieving a state of affairs where the agent will (hopefully correctly) have such a belief; for this, the plan will have a triggering event such as `+!served(dinner)`. Or, more generally, the agent might have to serve different kinds of meals, and in this case the triggering event would probably look like `+!served(Meal)` so that, if one particular event is that the agent has a new goal to have *dinner* served, then the plan would be executed with the logical variable `Meal` instantiated with `dinner`.

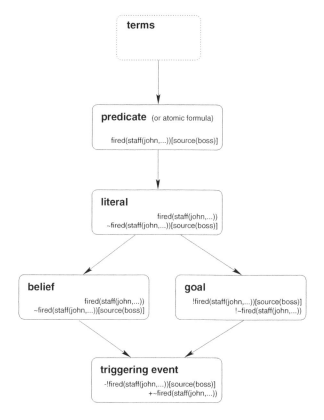

Figure 3.2 Types of AgentSpeak formulæ.

With regards to test goals, they are normally used to retrieve simple information from the belief base. For example, I normally remember how many packets of pasta I have in the larder, so when I need that information, I just retrieve it from memory. However, if I happen not to have that information to hand, I may need to perform an action specifically to find out this information. The action might be to ask someone else, or to go to the larder myself. This is why plans with test-goal addition as triggering events exist, even though with extra computing steps (and less transparency) this could be solved with achievement goals. (We will see later how this can be done.)

Plans with test goal addition in the triggering events exist because in some cases it might be useful to be able to write plans to retrieve information in a more elaborate way, by taking some action. When the agent has a test goal, the interpreter will first try to obtain the result for it by simply using facts (and possibly rules) in the belief base, but if that fails, then an event is generated so that a plan for that test goal might be executed, if possible. So, for the example above, we could write a plan with a triggering event +?stock(pasta,X) in order for the agent to obtain this information in case it happens not to be available in the agent's belief base when the information is needed for the execution of another plan.

Context

Recall that contexts are used to define when a plan should be considered *applicable*. Recall also that, because we are writing programs for dynamic environments, the idea is to postpone committing to a course of action until as late as possible. So, when it comes to choosing a plan – a course of action – in order to achieve a goal, it is the context that will be used to ensure that we are choosing a plan that is likely to succeed.

The way we write a plan context is as follows. It is often the case that a context is simply a conjunction of *default literals* and relational expressions – the syntax for expressions is very similar to Prolog, and also discussed below as part of the plan body. Recall that a literal is a predicate about some state of the world, which may have a strong negation. A default literal is a literal which may optionally have another type of negation, known in logic programming as *default negation*, which is denoted by the '**not**' operator. This is effectively used to check that a literal (with a strong negation or not) is not currently in the belief base. This gives us the four combinations of literals in Table 3.2. Logical expressions can appear in a plan context by combining, in the usual way, literals with the operators *and* (conjunction) denoted by '&' and *or* (disjunction) denoted by '|' as well as negation (the negation as failure operator 'not'). This exact same syntax can be used in the body of the rules that are part of the belief base.

Note that, for example, not believing that l is false (i.e. not believing $\sim l$) is not the same as believing l is true, because the agent might simply be ignorant about l (i.e. the agent does not know anything about l at all). This may sound a bit confusing, so a good thing to keep in mind is that, for those not already

Table 3.2   Types of literals in a plan context.

Syntax	Meaning
l	The agent believes l is true
$\sim l$	The agent believes l is false
not l	The agent does not believe l is true
not $\sim l$	The agent does not believe l is false

A plan context is typically a conjunction of these types of literals.

familiar with these notions, or for applications where it seems it might not be really
necessary to use this kind of representation, it may be better to use good old closed-
world assumption instead, and simply avoid using strong negation altogether.

According to what we have seen so far, examples of plan contexts could be as
follows:

```
+!prepare(Something)
   :   number_of_people(N) & stock(Something,S) & S > N
   <-   ... .


   ...


+!buy(Something)
   :   not ~legal(Something) & price(Something,P)
       & bank_balance(B) & B > P
   <-   ... .
```

That is, when the agent comes to have the goal of preparing something (to eat),
provided it has enough of it in stock for the number of people who will be eating,
the first plan can be used to handle that event (i.e. the 'event' that occurred because
there is a new goal to achieve). As we have not yet seen what the body of a plan
looks like, we gave just plan sketches above, hence the '...' to denote we have
omitted part of the plan.

The second outline plan above says that, when the agent comes to have a new
goal to buy something, provided it does not have explicit information saying it
is illegal to buy this thing, and it has enough money for the purchase, that plan
might be eligible for execution. Another example using strong negation is given
later, after we see the things we can write in a plan body.

One important thing to bear in mind is that any literal appearing in the context
is to be checked against the agent's *belief base*. For example, if stock(pasta,5) is
in the belief base, the result of literal stock(Something,S) appearing in a plan
context, assuming variable Something is already bound to pasta, will evaluate to
true with further instantiation of the plan variables, in particular S will become
bound to 5. This is generalised to checking whether the whole context is a logical
consequence of the belief base.

vanced

For the sake of completeness, it should be noted that the context can also have
formulæ that are not to be checked against the agent's beliefs. In particular,
there can be 'internal actions' which can be used, for example, to execute Java
code on the partial results already obtained in evaluating the context. Internal

actions can also appear in plan bodies, and their use in plan contexts is exactly the same. Therefore, we leave the use of internal actions to be explained in detail below, with the various other things that can appear in a plan body. However, at this point it is worth mentioning that programmers should be careful with internal actions in plan contexts, as they are executed *in order to check if the plan is applicable*; this means that, if a relevant event happens, those actions might be executed when the interpreter is determining the set of applicable plans, hence executed even if that plan turns out not to be the one selected to handle that event. Needless to say, an agent is likely to become very inefficient if heavy code is carelessly used in plan contexts.

The rules in the belief base can be used to make plan contexts much more compact; for example, instead of:

```
+!buy(Something)
    :  not ~legal(Something) & price(Something,P)
       & bank_balance(B) & B>P
    <-  ... .
```

we could add

```
can_afford(Something)
    :- price(Something,P) & bank_balance(B) & B>P.
```

to the belief base and then write the plan as:

```
+!buy(Something)
    :  not ~legal(Something) & can_afford(Something)
    <-  ... .
```

We now turn to the last part of a plan, its body.

Body

The body of a plan defines a course of action for the agent to take when an event that matches the plan's triggering event has happened and the context of the plan is true in accordance with the agent's beliefs (and the plan is chosen for execution). The course of action is represented by a sequence of formulæ, each separated from the other by ';'. There are six different types of formulæ that can appear in a plan body, and we now see each of them in turn.

Actions

One of the most important characteristics of an agent is that it must be able to *act* within an environment. Thus, unsurprisingly, a type of formula found in most plans is *actions*, which represent the things the agent is capable of doing. If we are programming a robot, we will know the actions the agent is capable of doing (those its hardware allow it to do). We will then need symbolic representations of these actions – which is how we will refer to them in the program – and then the overall agent architecture will interface these with the action effectors (i.e. the physical or software mechanism that executes the actions). We discuss this interface in Section 7.3, but for now, we only need to worry about the symbolic representation that will be used in the program.

The language construct for symbolically referring to such actions is simply a *ground predicate*. That is, we need to make sure that any variables used in the action become instantiated before the action is executed. One may wonder how the interpreter can differentiate actions from other uses of predicates. This is done simply by the position within the plan. For example, predicates in the context of the plan will be checked against the agent's beliefs, rather than being seen as actions. In the body of the plan, a simple predicate will be interpreted as an action: it will be clear below that all other types of formulæ that can appear in a plan body also have a predicate (or literal) involved but there are special notations to make it clear it is not an environment-changing action.

We will later see that there is another type of construct in the plan body called 'internal action' which, unlike the actions we are defining here, run internally within an agent rather than change the environment. We already saw an example of an internal action earlier: `.print(...)`, used to display text on a programmer's console.

As a simple example, suppose we are writing a program for a robot that has two arms ('right' and 'left') and the only action it can do is to rotate those arms – a rotation angle has to be specified. We could then have, in our plans, an action such as `rotate(left_arm,45)`, or we could choose to represent the actions as, say, `rotate_right_arm(90)` instead. Which we should choose is partly a matter of personal taste, and partly depends on the problem at hand.

Actions are executed by an agent's effectors, which are not part of the AgentSpeak interpreter. Therefore it makes sense to expect some kind of 'feedback' on whether the action was executed or not: it may turn out that it was impossible to carry out the action. Until such (Boolean) feedback is received, the plan that required the action to be executed must be *suspended* until we know the action has been executed. Executing an action does not mean that the expected changes will necessarily take place – this needs to be confirmed by perceiving the environment and checking subsequent beliefs. All the 'feedback' tells us is whether the requested action was executed at all; if it was not, the plan fails. We discuss this in more detail in the next chapter.

Achievement Goals

Having goals to achieve is another essential characteristic of an agent. Complex behaviour requires more than simple sequences of actions to execute; it involves achieving goals before further action can be taken. This is why we need to be able to say, as part of a plan body, that a goal has to be achieved, and only then the rest of the plan can be executed. As with all constructs we have seen so far, the representation will involve a literal; the way an achievement goal to be added to the agent's current goals is denoted is by prefixing the literal with the '!' operator. So, for example, we can say that the agent should have, as part of a course of action, the goal of achieving a state of affairs in which the customer is happy; for that, we could include in such plan body something like `!happy(customer)`, or to achieve a state of affairs in which the agent believes that there is no (longer any) gas leak, we could use `!~leaking(gas)`.

As we know from the types of triggering events we already saw, having a new goal is also considered an 'event' that can lead to the execution of a plan. This means that, if we include a goal in a plan body, a plan needs to be executed for that goal to be achieved, and only after the plan has successfully executed can the current course of action be resumed; so here again, the plan will be *suspended* (more on this in the next chapter). To be clearer, if we have '`a1; !g2; a3`' in a plan body, the agent will do `a1`, then it will have the goal of achieving `g2`, which will involve some other plan being executed, and only when that plan is finished can `a3` be executed. In fact, to be able to program proficiently in AgentSpeak, it helps to have a more detailed understanding of how exactly the interpreter works. This will be discussed in detail in Chapter 4.

A plan body is a *sequence* of formulæ to be executed. At times, it may be the case that we want the agent to have another goal to achieve but we do not need to wait until this has actually been achieved before we can carry on with a current course of action. The way to express this is to use the '`!!`' operator. In more advanced examples, we will see that this is also useful to make recursive plans more efficient. For the time being, just consider that, if the agent has '`!at(home); call(john)`' as part of a plan, it will only call John when it gets home, whereas if it is `!!at(home); call(john)`, the agent will call John soon after it comes to have a new goal of being at home, possibly even before it does any action in order to achieve that goal.

Test Goals

Test goals are normally used to retrieve information from the belief base, or to check if something we expected is indeed believed by the agent, during the execution of a plan body. One may wonder why we do not use the context of the plan to get all the necessary information before the agent starts on its course of action. Conceptually, the context should be used only to determine if the plan is applicable or not (i.e. whether the course of action has a chance of succeeding), and practically we may need to get the latest information which might only be available, or indeed might be more recently updated, when it is really necessary. For example, if the agent is guaranteed to have some information about a target's coordinates, which are constantly updated, and only needed before a last action in the course of action defined by the plan body, then it would make more sense to use `?coords(Target,X,Y);` just before that action than to include `coords(Target,X,Y)` in the context. In this particular example of test goal, presumably the variable `Target` would already be bound (by previous formulæ in the body) with the particular target in question, and assuming `X` and `Y` were free (i.e. not bound yet), any use of those variables after the test goal will be referring to the particular coordinates – as believed at the time the test goal is executed – of the target referred to by `Target`.

In the next chapter, we shall see that an agent's intentions are stacks of *partially instantiated plans*. This notion of instantiating variables with specific values as late as possible, as well as the late choice of a particular plan to achieve a goal, are very important for agent programming, precisely because of the dynamic nature of the environment.

As we mentioned in the description of triggering events, test goals can also be used for more complex tasks than just retrieving information already available in the belief base. Whilst we want to leave the instantiation of variables for as late as possible, there is an important issue with test goals. If the context of a plan does not follow from the agent's beliefs, the plan is not even selected to be executed. However, if we try to execute a plan which has a test goal in it and the test goal fails, then the whole plan where the test goal appeared fails, and the mechanism for handling plan failure will then be called upon. Therefore, care must be taken in the use of test goals.

This is partly the reason why creating plans with test-goal additions as triggering events might be interesting. Suppose we expect the agent to have some information while the plan is executed, so we introduce the appropriate test goal in the plan. If for some unanticipated reason the agent happens not to have the information required for the plan to be executed further, this would normally simply fail the plan. Instead, what happens in *Jason* is that, before failing, the interpreter first tries to create an event, which could match with a 'test-goal addition' triggering event in case the programmer created a plan to circumvent such situation; only in the case that *no* such (relevant) plan is available is the test goal effectively considered failed.

As a final note, consider that we could use an achievement goal to obtain the information required in the plan, as achievement goals can also be called with uninstantiated variables which become instantiated when the goal is achieved (i.e. a plan for it finishes executing). However, this forces the generation of an event for handling the achievement-goal addition, whereas a test goal normally (i.e. if the information is already in the belief base) just further instantiates the variables in the plan and the plan execution can be immediately resumed (if selected again in the next reasoning cycle, as we see in the next chapter). Note that, as we shall see in the next chapter, generating an event involves suspending the plan execution, finding a relevant plan, and so forth; this is just to emphasise that generating an event is rather expensive computationally, so it makes sense to have a specific test-goal construct.

Mental Notes

One important consideration when programming in *Jason* is to be aware of the things the agent will be able to perceive from the environment. Readers might be tempted to update an agent's beliefs about the consequences of its actions as part of the agent program, as they would in other action-based programming languages. This is *not* to be done in *Jason*. In *Jason*, some beliefs are acquired by perceiving the environment and those are updated automatically; we see in Section 7.2 how we can customise belief updates.

On the other hand, in practice it is often useful to be able to create beliefs during the execution of a plan. We call these beliefs *mental notes*, and they have a `source(self)` annotation to distinguish them within the belief base. An agent may need a mental note to remind itself of something it (or some other agent) has done in the past, or to have information about an interrupted task that may need to be resumed later, or a promise/commitment it has made. For example, consider an agent that needs to remember when the lawn was last mowed; this is done in the plan body with a formula `+mowed(lawn,Today);` or something similar. Recall that variable `Today` must have been instantiated previous to that. Then, when that formula appears in a plan being executed, a formula such as `mowed(lawn,date(2006,7,30))[source(self)]` will be added to the belief base.

A downside of this is that we normally have to explicitly delete mental notes that are no longer necessary, with a formula such as `-needs_be_done(mowing (lawn))`. The last operator in this category is available to simplify keeping the belief base tidy. Because in some cases we only need the last instance of a certain mental note in the belief base, the operator `-+` can be used to remove a former instance (if any exist) while adding the new one. For example, we could use `-+current_targets(NumTargets);` to keep only the current number of known targets (as instantiated to the `NumTargets` variable) and delete the information about what was previously believed to be the number of targets, in case any information about it was already in the belief base – note that attempting to delete a belief that does not exist is simply ignored, it does not cause a plan failure. Effectively, `-+current_targets(NumTargets);` is simply an abbreviation for `-current_targets(_); +current_targets(NumTargets);` where, as in Prolog, `_` is the anonymous variable (i.e. it can unify with any value, and is useful where the value is irrelevant for the rest of the plan).

If necessary, it is also possible to change beliefs that have another source rather than `self`, but care should be taken in doing so because this is rather unusual. In case it is necessary, one could always write `-trustworthy(ag2) [source(ag1)];` – which means that the programmer wants the agent to forget that agent `ag1` informed their agent that `ag2` is a trustworthy agent – or `+turned_on(tv)[source(percept)]` if the programmer insists in 'messing about' with the agent's perception of the environment.

Internal Actions

The first type of formula we saw were 'actions'. The important characteristic of such actions is that they *change the environment*; we can think of them as being executed *outside* the agent's mind (i.e. the agent's reasoner). It is often useful to think of some types of processing that the agent may need to do also as if it was an action being requested by the reasoner, except that it is executed within the agent; the whole processing of the action will be done as one step of the agent's reasoning cycle. This can be used by programmers to extend the programming language with operations that are not otherwise available, as well as providing easy and elegant access to legacy code – either in Java or other languages via the use of JNI (Java Native Interface).

The way an internal action is distinguished from (environment) actions is by having a '.' character within the action name. In fact, the '.' character is also used to separate the name of a library from the name of an individual internal action within that library (a Java 'package'). Suppose we are using *Jason* to program a controller for a robot, and we have a library for the digital image processing that the robot will require on the images captured by a camera it uses to sense the environment. The library is called `dip` and within it we have an internal action for finding a path (called `get_path`) to given coordinates from the robot's current location, in the last image taken by the robot's camera. In this case, we could

have, in some plan body, a formula such as `dip.get_path(X,Y,P)`; and variable
`P` would be instantiated by the internal action itself, with the path it found for the
robot to move to (`X,Y`) coordinates. We explain how we can create such internal
actions in Section 7.1.

Besides being able to create internal actions, it is very important that programmers familiarise themselves with *Jason*'s *standard internal actions* (a complete description of all currently available standard internal actions is given in
Appendix A.3). These are internal actions that are distributed with *Jason*, and
implement operations that can be very important for general BDI programming.
A standard internal action is denoted by an empty library name; that is, the action
symbol *starts* with the '`.`' operator. One particular standard action to bear in mind
already is the `.send` action, which is used for inter-agent communication. Using it
properly involves some knowledge of agent communication languages, which were
described in Section 2.5, and the particular communication language used in *Jason*
that we only see in detail in Chapter 6. For the time being, consider that this action
requires (at least) three parameters: the name of the agent that should receive the
message, the 'intention' that the sender has in sending the message (e.g. to inform,
to request information, to delegate goals) called the *performative*, and the actual
message content.

We can now show a simple example using some of the things we have already
seen about plans.

```
+!leave(home)
    :   not raining & not ~raining
    <-  !location(window);
        ?curtain_type(Curtains);
        open(Curtains);
        ....


+!leave(home)
    :   not raining & not ~raining
    <-  .send(mum,askIf,raining);
        ....
```

The first plan says that if the agent comes to have a new goal of leaving home in
a circumstance such that it is not sure whether it is raining or not — note that the
context of the plan says that the plan can only be used if the agent does *not* believe
it is raining but does *not* believe it is *not* raining either. In that circumstance, one
possible course of action to take is have the goal of being located at the window (a
plan to achieve this would require the agent to physically move in the environment,

if it was not already there), then retrieve from the belief base the particular type of curtains that the agent is now capable of perceiving, so that it can next open that particular type of curtains (presumably, different types of curtain require different types of action to get them open). The plan could then carry on, possibly by having the goal of checking whether an umbrella needs to be taken – we omit the rest of the plan for simplicity. Before the plan carries on, in the reasoning cycle after the `open(Curtains)` action was executed, the agent will have sensed the environment and therefore will have updated its own beliefs about whether it is raining or not.

The second plan offers an alternative course of action for exactly the same situation: having the goal of leaving home and not knowing whether it is raining or not. Instead of going to the window and opening the curtains, we simply ask the 'mum' agent, who (always) knows whether it is raining or not. Which plan is actually chosen for execution depends on the agent. (A 'teenager agent' will have the know-how defined by both plans, but in all likelihood would choose the second plan – possibly shouting rudely, but that is slightly more difficult to model here.) In practice, the choice of which *applicable plan* to use is done by one of the *selection functions*; this is explained in Chapter 4.

Expressions

Each formula in the context and body of a plan must have a boolean value (including internal actions). As in the short examples we saw when discussing plan contexts, it is often the case that we may need to use relational expressions, which also have boolean values. They can also be used in plan bodies, and the syntax is very similar to Prolog. For example, we can include in a plan context (resp. body) '`X >= Y*2;`' so that the plan is only applicable (resp. can only carry on executing) if this condition is satisfied. Note that, in a plan body, the plan will fail, in a similar way to when an action fails, if the condition is not satisfied.

The operators that can be used are listed in the grammar in Appendix A.1, and they are inspired by those available in Prolog. For readers not familiar with Prolog, note that `==` and `\==` are used for 'equal' and 'different', respectively – two terms need to be exactly the same for `==` to succeed. Another useful relational operator in logic programming is whether two terms can be *unified*; the `=` operator is used for this. Arithmetic expressions can be used in the usual way, both within relational expressions and within literals. One operator also available in Prolog that works slightly differently here (because of the predicate annotations not available in Prolog) is '`=..`', which is used to deconstruct a literal into a list. The resulting list has the format [⟨*functor*⟩, ⟨*list of arguments*⟩, ⟨*list of annotations*⟩], for example:
`p(b,c)[a1,a2] =.. [ p, [b,c], [a1,a2]].`

Annotations in triggering events and context

As we saw earlier, in *Jason* literals can have annotations, and literals appear in the triggering event and the context of a plan. This means that plans can make use of annotations for very interesting and useful things in multi-agent programs. In fact, this allows us to define more specific plans, which require particular annotations: omitting annotation effectively means that the plan is so general as to apply to any annotation. For example, suppose we want to write a plan for an agent that always accepts invitations to go down the pub, provided the invitation is from a good friend; we could have a plan such as:

```
+down_the_pub(Pub)[source(Agent)]
    :  good_friend(Agent)
    <- !location(Pub).
```

More typically, we will need to check the annotations we have introduced in the belief base in the plan context. For example, if we have annotated the beliefs with degrees of certainty (say, with a term degOfCert), we could have a plan such as:

```
+!contract(Agent)
    :  trustworthy(Agent)[degOfCert(C)] & C>0.9
    <-  ... .
```

If our agent is entering a contract with another agent, it might need different courses of action depending on how certain it already is that it can trust that other agent.

Higher order variables

In *Jason*, as in Prolog, we can bind variables to a predicate (or here, more generally, a formula) rather than a term. This allows us to do rather sophisticated things. For example, we could store an action formula as a term in the belief base, and use a variable to retrieve the action and execute it. Or we can use a variable to match *any* event, as in the example below – even though care should be taken as otherwise the agent might behave in unexpected ways as all events would be matched to that plan.

```
+!G[source(baddy)] : true <- true.
```

The plan tells the agent to ignore any goals that agent 'baddy' tries to delegate (in Chapter 6 we explain how agents delegate goals to one another). The keyword `true` used above can be used to denote an empty context, and also an empty body. Alternatively, we can omit those parts of a plan if they are empty. Thus the plan above could have been written simply '`+!G[source(baddy)].`' or we could have plans such as:

```
+!G[source(S)] : S \== goody.
```

omitting the body – the plan says that all goal delegations from agents other than 'goody' should be ignored; or

```
+B <- !~B.
```

omitting the context – this is a plan for a rebellious agent that will try to oppose everything it perceives in the environment. An example using variables bound to actions would be:

```
+B : simple_action_for(B,A) <- A.
```

In this case, obviously the programmer has to ensure that whatever terms A might be instantiated with is a valid agent action.

Plan labels

One thing we did not mention when explaining the plan syntax is that we can give a specific label to a plan; we can think of this as a 'name' for that plan. In fact, all plans have labels automatically generated by *Jason*, if we do not give specific names for each plan. This is important because, as we shall see later, it is sometimes necessary to be able to refer to a particular plan (instance) that the agent is running. Being able to specify labels is important not only because it might help us if we use meaningful names for the plans, but because we can also associate meta-level information about the plans within plan labels. The general notation for plans with a specific label is as follows:

```
@label  te : ctxt <- body.
```

However, the label itself does not have to be a simple term such as '`label`': it can have the same form of a predicate. That is, it can include annotations too, and it is in the annotations that we should include meta-level information (e.g. information that can help the interpreter choose one applicable plan over another), as exemplified below:

```
@shopping(1)[chance_of_success(0.7),usual_payoff(0.9),
    source(ag1),expires(autumn)]
+need(Something)
    :  can_afford(Something)
    <- !buy(Something).
```

and later (in Section 7.2) we see how we can write Java code for this meta-level information about the plan to be used.

Even though we can use a complex predicate, as above, as a label, it is normal practice to use a predicate of arity 0 (i.e. a proposition) but with annotations. For example, `@shopping1[...]` where `shopping1` would effectively be the name of the plan and all relevant meta-level information would be included in the annotations. There are pre-defined plan annotations which have special meaning for the interpreter; we discuss these in Chapter 4.4.

We have now seen all of the language constructs that can be used to program individual agents in *Jason*. The next section shows a complete agent program, which will clarify the role of these constructs and how they all come together.

3.4   Example: A Complete Agent Program

Consider the following example (inspired by [29]), which comes with *Jason*'s distribution.

> A domestic robot has the goal of serving beer to its owner. Its mission is quite simple, it just receives some beer requests from the owner, goes to the fridge, takes out a bottle of beer, and brings it back to the owner. However, the robot should also be concerned with the beer stock (and eventually order more beer using the supermarket's home delivery service) and some rules hard-wired into the robot by the Department of Health (in this example this rule defines the limit of daily beer consumption).

We use the Prometheus notation [75] in Figure 3.3 to give more details of the scenario. The system is composed of three agents: the robot, the owner and the supermarket. The possible perceptions that the agents have are:

- `at(robot,Place)`: to simplify the example, only two places are perceived, `fridge` (when the robot is in front of the fridge) and `owner` (when the robot is next to the owner). Thus, depending on its location in the house, the robot

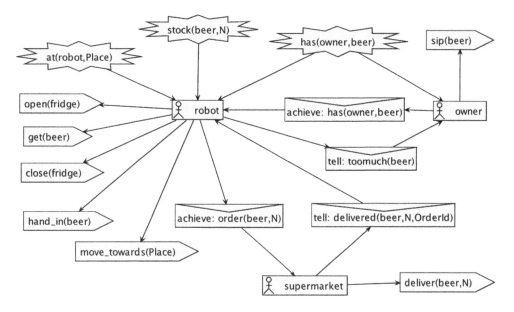

Figure 3.3 Overview of the domestic robot application.

will perceive either at(robot,fridge) or at(robot,owner), or of course no at percept at all (in case it is in neither of those places);

- stock(beer,N): when the fridge is open, the robot will perceive how many beers are stored in the fridge (the quantity is represented by the variable N);

- has(owner,beer): is perceived by the robot and the owner when the owner has a (non-empty) bottle of beer.

Most of the actions have obvious meanings, except perhaps move_towards. Again, to simplify the example, the way the robot moves in the house is greatly simplified. We assume that, when it performs the action move_towards(fridge), its hardware (or the simulator) performs one step towards the fridge, so the robot does not need to be concerned with the path to reach the fridge.

Besides perception and action, the agents also send messages to each other (messages are denoted by 'envelopes' in Figure 3.3). The owner sends a message to the robot when he wants another beer. The message type is 'achieve', so when the robot receives this message it will have has(owner,beer) as a new goal. A similar message is sent to the supermarket agent when the robot wants to order more beer. The supermarket sends a 'tell' message to robot when it has delivered the order. Tell messages are used to change the agent's beliefs (delivered(beer,N,OrderId) in this case). The AgentSpeak code for the agents in this application is as follows.

Owner Agent

```
!get(beer).  // initial goal

/* Plans */

@g
+!get(beer) : true
    <- .send(robot, achieve, has(owner,beer)).

@h1
+has(owner,beer) : true
    <- !drink(beer).
@h2
-has(owner,beer) : true
    <- !get(beer).

// while I have beer, sip
@d1
+!drink(beer) : has(owner,beer)
    <- sip(beer);
       !drink(beer).
@d2
+!drink(beer) : not has(owner,beer)
    <- true.

+msg(M)[source(Ag)] : true
    <- .print("Message from ",Ag,": ",M);
       -msg(M).
```

The owner's initial goal is `get(beer)`. Thus when it starts running the event `+!get(beer)` is generated and the plan labelled `@g` is triggered (most plans in the source are identified by a label prefixed by '@'). This plan simply sends a message to the robot asking it to achieve the goal `has(owner,beer)`.

When the robot has achieved the goal, the owner perceives this, and the belief `has(owner,beer)[source(percept)]` is added to the belief base. This change in the belief base generates an event that triggers plan `@h1`. When executing `@h1`, the subgoal `!drink(beer)` is created, and then the owner starts drinking, by recursively executing plan `@d1`. The event `+!drink(beer)` has two relevant plans (`@d1` and `@d2`) and they are selected based on their context: `@d1` is selected whenever the owner has any beer left and `@d2` otherwise.

The beer will eventually be finished (the percept has(owner,beer) is removed), and the intention of drinking also finishes (the !drink(beer) plan recursion stops). The event -has(owner,beer) then triggers plan @h2. This plan creates a new subgoal (!get(beer)), which starts the owner behaviour from the beginning again.

The last plan simply prints a special type of message received from an agent identified in the source as Ag. Note that the message is removed from the belief base after the .print internal action is executed. This is necessary because, when such message arrives, an event is generated only in the case where that message is not already in the belief base.

Supermarket Agent

```
last_order_id(1). // initial belief

// plan to achieve the the goal "order" from agent Ag
+!order(Product,Qtd)[source(Ag)] : true
  <- ?last_order_id(N);
     OrderId = N + 1;
     -+last_order_id(OrderId);
     deliver(Product,Qtd);
     .send(Ag, tell, delivered(Product,Qtd,OrderId)).
```

The code of the supermarket agent is quite simple. It has an initial belief representing the last order number and a plan to achieve the goal of processing an order for some agent. The triggering event of this plan has three variables: the product name, the quantity and the source of the achievement goal (the sender of the achieve message in this case). When selected for execution, the plan consults the belief base for the last order number (?last_order_id(N)); increments this number (OrderId = N + 1); updates the belief base with this new last order number (-+last_order_id(OrderId)); delivers the product (through an *external* action); and finally sends a message to the client (the robot, in this case) telling it that the product has been delivered.

Robot Agent

```
/* Initial beliefs */

// initially, I believe that there are some beers in the fridge
available(beer,fridge).

// my owner should not consume more than 10 beers a day :-)
```

```
limit(beer,10).

/* Rules */

too_much(B) :-
   .date(YY,MM,DD) &
   .count(consumed(YY,MM,DD,_,_,_,B),QtdB) &
   limit(B,Limit) &
   QtdB > Limit.

/* Plans */

@h1
+!has(owner,beer)
   :  available(beer,fridge) & not too_much(beer)
   <- !at(robot,fridge);
      open(fridge);
      get(beer);
      close(fridge);
      !at(robot,owner);
      hand_in(beer);
      // remember that another beer will be consumed
      .date(YY,MM,DD); .time(HH,NN,SS);
      +consumed(YY,MM,DD,HH,NN,SS,beer).

@h2
+!has(owner,beer)
   :  not available(beer,fridge)
   <- .send(supermarket, achieve, order(beer,5));
      !at(robot,fridge). // go to fridge and wait there.

@h3
+!has(owner,beer)
   :  too_much(beer) & limit(beer,L)
   <- .concat("The Department of Health does not allow me ",
              "to give you more than ", L,
              " beers a day! I am very sorry about that!",M);
      .send(owner,tell,msg(M)).

@m1
+!at(robot,P) : at(robot,P) <- true.
```

```
@m2
+!at(robot,P) : not at(robot,P)
  <- move_towards(P);
     !at(robot,P).


// when the supermarket finishes the order, try the 'has'
// goal again
@a1
+delivered(beer,Qtd,OrderId)[source(supermarket)] : true
  <- +available(beer,fridge);
     !has(owner,beer).


// when the fridge is openned, the beer stock is perceived
// and thus the available belief is updated
@a2
+stock(beer,0)
   :  available(beer,fridge)
   <- -available(beer,fridge).


@a3
+stock(beer,N)
   :  N > 0 & not available(beer,fridge)
   <- +available(beer,fridge).
```

The robot is the main agent in this example, and its code is thus more complex. The robot maintains a belief regarding availability of beer in the fridge: initially, it believes `available(beer,fridge)`. It is important for the robot to remember whether or not there is beer available irrespective of its location in the environment. The robot does not perceive the beer everywhere: the stock is only perceived when the robot is in front of the fridge and it is open. The `stock` perception is removed from the belief base when the robot, for instance, closes the fridge. Thus, it needs to remember what it saw inside the fridge. Plans `@a2` and `@a3` update the availability belief according to the stock perception.

The belief base also has a rule to determine whether the owner already drank the maximum number of beers allowed per day. This verification is based on the number of `consumed` beliefs. When the robot gives a beer to the owner it adds a belief to remember that; this belief has the form `+consumed(Year,Month,Day, Hour,Minute,Second,beer)` (see the end of plan `@h1`). The rule gets the current date using the internal action `.date`; then uses another internal action (`.count`) to count how many `consumed` beliefs there are in the belief base (the variable `QtdB` unifies with this quantity). The `too_much` belief holds if `QtdB` is less than the limit of beers (10 in this example).

The main task of the robot is to achieve the goal `!has(owner,beer)`, created by an achieve message received from the owner. The robot has three plans to achieve this goal (`@h1`, `@h2`, and `@h3`). The first plan is selected when there is beer available and the owner has not drunk too much beer today (see the plan context). In this plan execution, the robot goes to the fridge (subgoal), opens it (external action), gets a beer (external action), closes the fridge (external action), goes back to the owner (subgoal) and finally gives the beer to the owner (external action). The plans to achieve the `!at` subgoal are quite simple and use the `move_towards` external action until the agent believes it is in the target place `P`.

The second plan is selected when there is no more beer in the fridge. In this case, the robot asks the supermarket for five more bottles of beer, goes to the fridge, and waits for the beer delivery. This plan finishes when the robot reaches the fridge. However the `!has(owner,beer)` goal should be resumed when the beers are delivered. Plan `@a1` does that when a message from the supermarket is received.

The third plan is (unfortunately) selected when the robot is forbidden to give more beers to the owner. In this case, it says a special message to the owner explaining the reasons.

Running the System

Every system developed with *Jason* has a multi-agent system definition file where the set of participating agents, the shared environment and various parameters for the multi-agent system execution are defined. The various configuration options that can be used in the multi-agent system definition file will be presented throughout the book, and Appendix A.2 has a complete list of such options; for now it is enough to define the three agents that compose the system and the Java class that simulates the environment:[1]

```
MAS domestic_robot {

    environment: HouseEnv(gui)

    agents: robot;
            owner;
            supermarket agentArchClass SupermarketArch;
}
```

---

[1]The class that implements the environment is **houseEnv** and is explained in Example 5.2.

An excerpt of the result of the system execution is:

```
[robot] doing: move_towards(fridge)
[robot] doing: move_towards(fridge)
[robot] doing: move_towards(fridge)
[robot] doing: open(fridge)
[robot] doing: get(beer)
[robot] doing: close(fridge)
[robot] doing: move_towards(owner)
[robot] doing: move_towards(owner)
...
[robot] doing: hand_in(beer)
[owner] doing: sip(beer)
[owner] doing: sip(beer)
...
[supermarket] doing: deliver(beer,5) ...
[owner] saying: Message from robot: The Department of
Health does not allow me to give you more than 10 beers
a day! I am very sorry about that!
```

When running the examples available with the *Jason* distribution, and also when developing your own systems, note that multi-agent systems definition files have an extension .mas2j (for historical reasons) and files with AgentSpeak code have an extension .asl.

3.5 Exercises

Basic
1. The robot used in Example 3.4 has plans (identified by h1-h3) triggered when the agent has a goal that unifies has(owner,beer) despite the source of the goal (i.e. personal or delegated). Change these plans to be triggered only if the source of the goal is the owner agent.

Basic
2. Improve the code of the Supermarket agent of the example so that it manages its stock. It initially has, for instance, 100 bottles of beer and this value will decrease as it delivers beer to the robot. Of course, it can only deliver beer when it has enough beer in its stock, otherwise it should inform the robot that it has no more beer in its stock.

Basic
3. The robot asks the supermarket for more beer only when the fridge is empty (see +stock plans). Modify the robot code so that its behaviour is more pro-active and it asks for more beer before it is finished. Note that: (i) it is not a case of simply changing 0 to 1 in plan @a2, because even with only 1 beer remaining in the fridge, there is still beer available, so the belief about beer being available should not be removed; (ii) plan @h2 may be removed from

the code because the robot is now pro-active and there will always be beer available; (iii) when some beer is delivered (see plan @a1), the robot does not need to achieve the goal has(owner,beer) anymore.

4. In the robot code, we have two plans for the goal !at. What happens if we change their order (m2 before m1)? What happens if the second plan does not have a context, as in the following code? `Basic`

```
@m1 +!at(robot,P) : at(robot,P) <- true.
@m2 +!at(robot,P) : true
        <- move_towards(P);
           !at(robot,P).
```

Is the behaviour of the robot the same as with the original code? In this case, what happens if we change the order of the plans?

5. Create a new supermarket agent, initially with the same code as the supermarket of the example. Change the code of both supermarkets such that, once they have started, they send the following message to the robot, informing of their price for beer. `Advance`

```
.send(robot,tell,price(beer,3))
```

The robot should then buy beer from the cheapest supermarket. To help the robot coding, write a (Prolog-like) rule that finds the cheapest supermarket according to the robot's belief base. Note that the robot will always have two beliefs of price from different sources (the two supermarkets).

4

Jason Interpreter

4.1   The Reasoning Cycle

We now describe how the *Jason* interpreter runs an agent program. An agent operates by means of a *reasoning cycle*, which in the case of *Jason* we can divide into 10 main steps. Figure 4.1 shows the architecture of a *Jason* agent as well as the component functions that are executed during the reasoning cycle. We can think of the reasoning cycle as being somewhat analogous to the BDI decision loop that we saw in Chapter 2.

In Figure 4.1, rectangles represent the main architectural components that determine the *agent state* (i.e. the *belief base*, the *set of events*, the *plan library*, and the *set of intentions*); rounded boxes, diamonds and circles represent the functions used in the reasoning cycle. Both rounded boxes and diamonds represent the functions that can be customised by programmers, whereas circles are essential parts of the interpreter that cannot be modified. The difference between rounded boxes and diamonds is that the latter denote *selection functions*: they take a list of items and the function selects one of them.

The AgentSpeak program determines the initial state of the belief base, the set of events, and the plan library. The beliefs in the program are used to initialise the belief base; such initialisation generates belief additions, so these belief addition events go in the initial set of events; initial goals also go as events to the set of events; and the plans in the program code form the (initial) plan library. When the agent starts running, the set of intentions is empty.

We explain in detail each of the 10 steps of the reasoning cycle in turn. Note that some elements of Figure 4.1 are labeled with numbers, which refer to steps of the reasoning cycle as presented below.

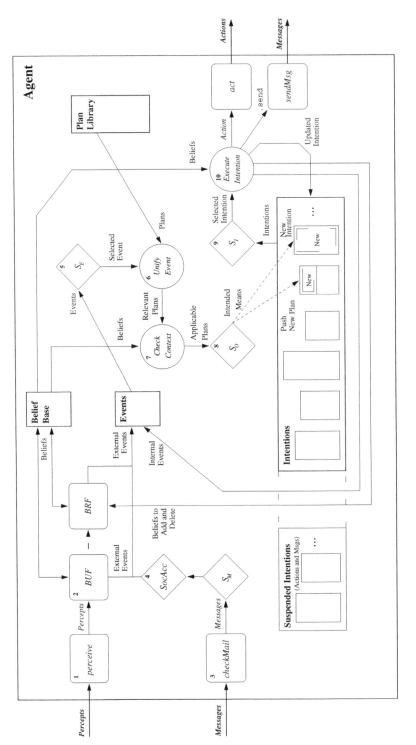

Figure 4.1 The *Jason* reasoning cycle.

Step 1 – Perceiving the Environment

The first thing an agent does within a reasoning cycle is to sense the environment so as to update its beliefs about the state of environment. The overall agent architecture must have a component that is capable of perceiving the environment, in a symbolic form as a list of literals. Each literal is a *percept*, which is a symbolic representation of a particular property of the current state of the environment.

The perceive method is used to implement the process of obtaining such percepts. In the default implementation available in *Jason*, the method will retrieve the list of literals from a simulated environment implemented in Java – Chapter 5 explains the support available in *Jason* for developing Java code for simulated environments. In an application where, for example, the agent will have access to sensor data from real-world devices or systems, the perceive method will need to interface with such devices.

vanced

One of the things that can be done using the support for programming environments is to determine properties of the environment which are only perceivable by one (or some) particular agent(s); that is, *individualised perception* can be defined. On top of that, in certain applications, programmers may want to have an agent dealing with faulty perception of the simulated environment. This is another reason for customisation of the perceive method. That is, something like a filter of percepts can be implemented within this method, for example to simulate faulty perception. Customisation of this method is explained in Section 7.2.

Step 2 – Updating the Belief Base

Once the list of percepts has been obtained, the belief base needs to be updated to reflect perceived changes to the environment. This is done by a *belief update function* and the (customisable) method implementing it is called buf. The default buf method assumes that everything that is currently perceivable in the environment will be included in the list of percepts obtained in the previous step; that is, when the agent senses the environment, it perceives everything that is currently perceptible to itself (much of which the agent might already have perceived). Therefore, all the buf method does is to update the belief base in the following very simple way (where p is the set of current percepts and b is the set of literals in the belief base that were obtained from sensing the environment last time around):

1. each literal l in p not currently in b is added to b;

2. each literal l in b no longer in p is deleted from b.

Each change in the belief base during the execution of the belief update function generates an *event*. In particular, the type of event that is generated because beliefs are added or deleted from the belief base in updates resulting from perception of the environment is called an *external event*. We will see later that internal events have an intention associated with them – the intention that caused the generation of the event. So, an event is effectively represented as a pair, the first component denoting the change that took place, and the second being the associated intention. As external events are not generated by intentions, they are represented by having an *empty intention* associated with them; the empty intention is denoted by ⊤.

To give an example, consider an agent that is perceiving the colour of a box (identified as box1) for the first time. It will then acquire a new belief such as colour(box1,red)[source(percept)] and the event that is generated because of this would be:

$$\langle +\texttt{colour(box1,red)[source(percept)]}, \top \rangle$$

As a consequence, a plan relevant for +colour(box1,red) might be executed. Note also that if the box is still within the agent's sensory range in the next reasoning cycle(s), no *change* will have occurred, and hence *no event is generated*. On the other hand, if at some future point the agent comes not to perceive that box (or rather its colour) any longer, then the following event is generated:

$$\langle -\texttt{colour(box1,red)[source(percept)]}, \top \rangle$$

This event might also trigger a plan. If later the box is perceived again, another addition event would take place, and as a consequence a plan relevant for +colour(box1,red) might be executed again.

Another thing to bear in mind is with respect to how annotations influence event generation due to belief changes. In general terms, if a belief $b[a_1,a_2]$ is added to the belief base when it already contains that belief with some annotations, as in $b[a_1,a_3,a_4]$, the actual event generated is $+b[a_2]$, as this is the only change with respect to the current state of the belief base.

Advanc

As noted above, what matters are *changes* in what the agent can perceive in the environment. Given that only changes are important, it is not difficult to realise that receiving a list of *all* percepts and comparing those with the belief base is not very efficient. For large applications, programmers will want to customise the **perceive** method, e.g. so that it is only notified of actual changes in the environment. Also, as discussed in [2], sophisticated agents may need more elaborate belief revision functions (brf), and it will probably be the case that the belief update method (buf) would need to let the belief revision function take care of the actual changes in the belief base (this is what is meant by the dotted arrow from buf to brf in Figure 4.1).

Step 3 – Receiving Communication from Other Agents

Other important sources of information for an agent in a multi-agent system are other agents in the same system. At this stage of the reasoning cycle, the interpreter checks for messages that might have been delivered (by the distribution infrastructure being used within the *Jason* platform) to the agent's 'mailbox'. This is done by the checkMail method, which again can be customised (as explained in Section 7.3). This method simply gets the received messages stored by the underlying multi-agent system infrastructure and makes them available at the level of the AgentSpeak interpreter.

Within one reasoning cycle, only *one* message is processed by the AgentSpeak interpreter (used by *Jason*). In many scenarios, agents will want to give priority to certain messages, and this is why we need an agent-specific *selection function* to allow such prioritisation. We will see that other such user-defined selection functions are used by the interpreter in later steps. The selection functions are implemented as methods that can be overridden by the programmer; this type of *Jason* customisation is explained in Section 7.2. The *message selection function* (S_M) selects, among all previously received messages yet to be processed, the one that will be processed in this reasoning cycle. Its default implementation in *Jason* simply chooses the first message in the list; that is, unless redefined by the programmer, *Jason* will process communication messages in the order in which they were received.

Step 4 – Selecting 'Socially Acceptable' Messages

Before messages are processed, they go through a selection process to determine whether they can be accepted by the agent or not. The interpreter function that does this is called *social acceptance function* and is implemented by a method called SocAcc. This method will normally need customising, possibly for each individual agent. A particular implementation of the method determines which agents can provide information and know-how, as well as delegate goals, to the particular agent using that implementation of the SocAcc method. The default implementation simply accepts all messages from all agents; the customisation of this method is explained in Section 7.2.

Note that cognitive reasoning about how to handle messages should be implemented in the AgentSpeak program, rather than the SocAcc method. For example, once the agent has information in its belief base about information sent from other agents, it can choose to use that or not based on how much the sender is trusted; or once a goal has been delegated, the course of action the agent will take, if any, can be determined for example through the context of the relevant plans (and we later see that this can also involve the agent checking its other current intentions, etc.). The purpose of the SocAcc method is to avoid receiving communication from agents with whom an agent should have no relationship (think of it as a kind of spam filter!).

In particular, agent-oriented software engineering methodologies which account for the organisational level of multi-agent systems will certainly provide the means to determine the kinds of relationships that agents within an agent society can be expected to have. Such an organisational design of a multi-agent system can help determine the social relationships that need to be allowed in the SocAcc method implementation. There is ongoing work on combining one particular organisational model called $\mathcal{M}oise^+$ with *Jason*; see Section 11.2 for further discussion.

Handling messages generally implies updating the belief base, the set of events or the plan library. We shall see why in Chapter 6, where we also see that the way the semantics of communication is implemented in *Jason* is, interestingly, done by pre-loaded AgentSpeak plans, each dealing with a specific performative. Therefore, if we need to update the belief base to include some information sent by another agent, this will be actually done by an AgentSpeak plan. In order to get such a plan executing (so that the update can be done), an (external) event needs to be generated (similarly as discussed in Step 2 of the reasoning cycle), hence the arrow from the SocAcc function to the set of events in Figure 4.1.

Steps 1–4 were all concerned with the agent updating its beliefs about the world and other agents. Steps 5–10, below, describe the main steps of the interpretation of AgentSpeak programs. The process starts with the belief base and the set of events having been updated and then goes on with two other important parts: one of the events is selected to be handled in the current reasoning cycle (leading to a new or updated intention), and one of the agent's intentions is further executed.

Step 5 – Selecting an Event

Practical BDI agents operate by continuously handling *events*, which represent either perceived changes in the environment or changes in the agent's own goals.

In each reasoning cycle, only one pending event will be dealt with. There can be various pending events, for example because various different aspects of the environment changed recently and the agent has not gone through enough reasoning cycles to handle them all. Therefore, we need to *select* an event to be handled in a particular reasoning cycle. This is done by an agent-specific *event selection function* ($\mathcal{S}_\mathcal{E}$), which may also be customised.

Normally, this selection function will need to be customised to account for any priorities that are specific for each agent in each application. Only in simple applications, where all events can be assumed to have the same importance for the agent, would the programmer be able to use the predefined event selection function. The 'set of events' is implemented in *Jason* as a list, and new events are added to the end of the list. What the default implementation of the event selection function does is simply to select the first event in the list. If the programmer chooses *not* to customise the event selection function, the **Events** component in Figure 4.1 can therefore be thought of as a *queue* of events (i.e. a FIFO structure). Section 7.2 explains how selection functions may be customised.

To give an example, suppose there are currently two events in the set of events: first ⟨+colour(box1,blue)[source(percept)], ⊤⟩ and second ⟨+colour(sphere2,red)[source(percept)], ⊤⟩. If no customised event selection function has been defined for this agent, the event for the newly perceived blue box will be the one to be handled in this reasoning cycle. However, if for this agent red objects are of special interest and should be handled before any action is taken regarding objects of other colours, and the appropriate selection function implementation has been provided, then the second event will be considered the *selected event* for this cycle. As we shall see, the next steps of the reasoning cycle consider that there is a selected event; the selected event is *removed* from the set of events. If the set of events turns out to be empty (i.e. there have been no changes in the agent's beliefs and goals since the last event was handled), then the reasoning cycle proceeds directly to step 9.

Step 6 – Retrieving all Relevant Plans

Now that we have a selected event, we need to find a plan that will allow the agent to act so as to handle that event. The first thing to do is to find, in the **Plan Library** component (see Figure 4.1) all plans which are *relevant* for the given event. This is done by retrieving all plans from the agent's *plan library* that have a triggering event that can be *unified* with the selected event; the particular form of unification (including annotations) used here will be informally explained with a short example (see Chapter 10 for a more formal account). Let us assume that the selected event was ⟨+colour(box1,blue)[source(percept)], ⊤⟩ and

these are all the plans the agent has in its plan library (we are only interested in the triggering event part of the plans at this point, hence the ' ... '
omissions):

```
@p1   +position(Object,Coords) :  ...  <-  ...  .
@p2   +colour(Object,Colour) :  ...  <-  ...  .
@p3   +colour(Object,Colour) :  ...  <-  ...  .
@p4   +colour(Object,red) :  ...  <-  ...  .
@p5   +colour(Object,Colour)[source(self)] :  ...  <-  ...  .
@p6   +colour(Object,blue)[source(percept)] :  ...  <-  ...  .
```

Comparing the predicate symbol of the triggering event of plan p1 (i.e. position),
we can see it does not match the predicate symbol (colour) of the first component of the event (for the time being, ignore the second component of the event).
This means we can already discard plan p1 as it is certainly not relevant for the
selected event. If we try to unify colour(box1,blue) that appears in the event
with colour(Object,Colour) in the triggering events of plans p2 and p3 we
succeed with a unifying function that maps variable Object to term box1 and
Colour to blue; that is, in both cases (plans p2 and p3) the unifying function
is {Object ↦ box1, Colour ↦ blue}. Plan p4 in turn is not relevant because in
attempting to unify, term red in the triggering event does not match term blue in
the event. That is, plan p4 is only relevant for red objects.

For plan p5 to be relevant, we would need to find source(self) within the
list of annotations of the actual event. As the only annotation in the event is
source(percept), plan p5 is not considered relevant. The intuition is that this
plan is very specific: it is only relevant for information about colour that has
been added to the belief base by the agent itself as a mental note, which is not
the case for this event. Finally, the predicate in the triggering event of plan p6
unifies with the event via the substitution {Object ↦ box1}, and the annotation
source(percept) in the plan *is* a member of the annotations of the actual event,
making p6 another relevant plan. Therefore, in this running example, at the end
of step 6 we would have determined that the *set of relevant plans* for the selected
event is {p2, p3, p6}, and we can then go on to the next step where we will further
select, from within this set, the *set of applicable plans*.

Advance

There are some further considerations with respect to selecting relevant plans.
If, after attempting to unify an *external event* with the triggering event of all
plans available in the plan library, the set of relevant plans remains empty, the
event is simply *discarded*. Note that, in typical environments, an agent will
perceive various things which may be useful information but do not prompt

the agent to act upon them or to have new goals to achieve. Therefore, it is reasonable that external events which are not relevant for the agent are simply discarded. As can be concluded from this, all the changes in beliefs which may require the agent to act, or to revise its current planned actions, need to appear in the triggering events of the plans that will be included the agent's plan library. We postpone until Section 4.2 the discussion of what happens if there are no relevant plans for internal events (as this will result in plan failure); in Section 4.3 we explain how programmers can configure the interpreter to 'requeue' rather than discard events in some circumstances.

Step 7 – Determining the Applicable Plans

We saw earlier that plans have a *context* part which tells us whether a plan can be used at a particular moment in time, given the information the agent currently has. Thus, even though we have selected all the *relevant* plans, we cannot just use any of them to be the particular course of action the agent will commit to take so as to handle the selected event. We need to select, from the relevant plans, all those which are currently *applicable*; that is, we try to use a plan which, given the agent's know-how and its current beliefs, appears to have a chance in succeeding. In order to do this, we need to check whether the context of each of the relevant plans is believed to be true; in other words, whether the context is a *logical consequence* of the agent's belief base. To give an informal understanding of the notion of logical consequence, including *Jason* annotations, consider the following examples (Definition 10.1 provides a formal account). Suppose the belief base is currently as follows:

```
shape(box1,box)[source(percept)].
pos(box1,coord(9,9))[source(percept)].
colour(box1,blue)[source(percept)].
shape(sphere2,sphere)[source(percept)].
pos(sphere2,coord(7,7))[source(bob)].
colour(sphere2,red)[source(percept),source(john)].
```

and we now also show the contexts of those plans that, as we saw earlier, were included in the set of relevant plans:

```
@p2
+colour(Object,Colour)
    :   shape(Object,box) & not pos(Object,coord(0,0))
    <-  ...  .
```

```
@p3
+colour(Object,Colour)
    :   colour(OtherObj,red)[source(S)] & S\==percept &
        shape(OtherObj,Shape) & shape(Object,Shape)
    <-  ...  .

@p6
+colour(Object,blue)[source(percept)]
    :   colour(OtherObj,red)[source(percept)] &
        shape(OtherObj,sphere)
    <-  ...  .
```

Recall that, for plans p2 and p3, the unifying function is already {Object ↦ box1, Colour ↦ blue}. Therefore, to satisfy the first part of the conjunction in plan p2's context, we need to be able to find, within the belief base, the predicate shape(Object,box) with Object substituted by box1. As we have shape(box1,box) in the belief base, that part of the context is satisfied. Now we need to make sure that pos(box1,coord(0,0)) *cannot* be found in the belief base, which is also the case, hence plan p2, besides being relevant, is in the *set of applicable plans* as well.

Now we need to check whether plan p3 is also applicable. The first formula in the context has variables OtherObj and S, both as yet uninstantiated, so we obtain values for them by searching for an object for which the colour is believed to be red: in this case, the first attempt will be to combine the previous unifying function with {OtherObj ↦ sphere2, S ↦ percept}. Note that there are two different sources for that information, so this is not the only possibility for unification that satisfies the first predicate in the context. The second formula in the context requires S, now substituted to percept, to be different from percept, so with this unifying substitution we cannot satisfy the context. However, we did have an alternative which we must now try, namely {OtherObj ↦ sphere2, S ↦ john}. This latter substitution of variables *does* satisfy the second formula in the context (which requires the source of whatever red object we found not to be perceptual). The third formula then tells us to get variable Shape instantiated with whatever the agent believes to be the shape of OtherObj (currently being substituted by sphere2) – in this case we will have Shape mapping to sphere. However, the fourth formula now requires us to find (after the appropriate substitutions) shape(box1,sphere) in the belief base, which does *not* succeed. Reading the third and fourth formulæ together, what they are telling us is that whatever other (red) object we found needs to have the same shape of the object whose colour we just perceived, if this particular plan is to be used for execution. The current substitutions then did not succeed in satisfying the context of plan p3. As we do not have any further alternative instantiations given the current state of the belief base, plan p3 is *not* in the set of applicable plans.

Finally, checking the context of plan p6 requires us to find any red object, provided the belief about the object's colour was obtained by sensing the environment, and then we have to check whether that object is a sphere. In other words, this plan for dealing with the event of the agent perceiving a new blue object is only *applicable* if the agent already has beliefs about a red sphere as well. In our agent's current situation, this is satisfied by combining {OtherObj ↦ sphere2} with the existing unifying function for that plan. In summary, the set of applicable plans is {p2, p6}, which means we have both p2 and p6 as alternative ways to handle the selected event given the agent's current belief state. Having more than one plan in the set of applicable plans means that, given the agent's know-how and its current beliefs about the world, any of those plans would be appropriate for handling the event (i.e. would successfully achieve a goal, or cope with perceived changes in the environment). The next step of the reasoning cycle deals with choosing one of those alternative courses of action.

---

vanced

Note that the unifying function that was found while checking if a plan is currently applicable is very important, as it will apply to the whole plan. That is to say that, if any of the variables that appeared in the head of the plan also appears in the body, they will be substituted by the value which the unifying functions determined in this step attributed to them. Therefore, we need to keep track of these unifying functions. This is why an actual set of applicable plans would look something like:

$$
\begin{aligned}
&\{ \\
&\quad \langle \text{p2}, \{\text{Object} \mapsto \text{box1}, \text{Colour} \mapsto \text{blue}\}\rangle, \\
&\quad \langle \text{p6}, \{\text{Object} \mapsto \text{box1}, \text{OtherObj} \mapsto \text{sphere2}\}\rangle \\
&\}
\end{aligned}
$$

As in one of the examples above, there are often various different ways of substituting variables for terms so that the context can be satisfied. In the case of plan p3 above, there were only two and neither were satisfactory. However, in most cases when a plan is applicable at all, there will be various different (mostly general) unifiers, all of which make the plan applicable. For example, for plan p6 the agent could have perceived various different red spheres, all of which would satisfy the context. In some cases one may want the next step of the reasoning cycle to choose not only between different applicable plans but also between different substitutions for the applicable plans. We explain how programmers can take advantage of this in Section 4.4.

Step 8 – Selecting One Applicable Plan

Given the agent's know-how as expressed by its plan library, and its current information about the world as expressed by its belief base, we have just determined that all the plans currently in the set of applicable plan are suitable alternatives to handle the selected event. This means that, as far as the agent can tell, any of those plans are currently suitable in the sense that executing *any* of them will 'hopefully' suffice for dealing with the particular event selected in this reasoning cycle. Therefore the agent needs to chose *one* of those applicable plans and commit to executing that particular plan. Committing to execute a plan effectively means that the agent has the *intention* of pursuing the course of action determined by that plan, so one can expect that the selected plan will soon 'end up' in the agent's **Set of Intentions** (the large box in Figure 4.1). The chosen applicable plan is called *intended means* because the course of action defined by that plan will be the means the agent intends (or is committed) to use for handling the event – recall that an event is normally either a goal to achieve or a change in the environment to which the agent needs to react.

The selection of the particular plan from within the set of applicable plans that will be included in the set of intentions is done by another selection function, called *option selection function* or *applicable plan selection function* (S_O). Each of the applicable plans can be considered one of various *options* the agent has in regards to viable alternatives for handling the selected event. Note further that these are alternative *courses of actions* for handling *one particular event*, where an event represents one particular goal or one particular change of interest perceived in the environment. Note that the goals currently in the set of events represent different desires the agent can choose to commit to, while different applicable plans for one such goal represent alternative courses of action the agent can use to achieve that particular goal.

As with the other selection functions, S_O is also customisable, as explained in Section 7.2. The pre-defined option selection function chooses an applicable plan based on the order in which they appear in the plan library. This in turn is determined by the order in which plans are written in an agent source code, or indeed the order in which plans are communicated to the agent (exchange of plans through communication is discussed in Chapter 6). This order can be useful for example if a triggering event requires recursive plans, in which case the programmer knows that the plan for the end of the recursion should appear first, as is standard practice in Prolog.

In Figure 4.1, one can see that there are two (dashed) lines from the S_O selection function to the set of intentions. This is because there are two different ways of updating the set of intentions, depending on whether the selected event was internal or external. Recall that internal events are changes in goals and external events are perceived changes in the environment.[1] If the agent acquires a new intended means (the selected applicable plan) because some change was noticed in the environment, that will create a *new intention* for the agent. Each separate intention within the set of intentions therefore represents a different *focus of attention* for the agent. For example, if a person hears the door bell and the mobile phone ringing at the same time, they could commit to handling both events, for example by communicating over the phone while moving towards the door. Similarly, an agent might commit to handle different external events at the same time, and all the respective intentions will be competing for the agent's attention. In the next step we see how one particular intention (one of the focuses of attention) is chosen for further execution at each reasoning cycle.

On the other hand, internal events are created when, as part of an intended means, the agent gets to have a new goal to achieve. This means that, before the course of action that generated the event (i.e. a goal addition) is resumed, we need to find, and execute to completion, a plan to achieve that goal. Therefore, for internal events, instead of creating new intentions, we push another intended means on top of an existing intention. This forms a *stack* of plans (for a given intention/focus of attention), which is very convenient because the interpreter knows that it is the plan at the top of the intention that *may* currently be executed (subject to competing with other focuses of attention). To make this clearer, consider the following example. For the issue of what is done with a new intended means, an abstract example should be sufficiently clear, so we use b to represent beliefs, p to represent plans, g to represent goals, and a to represent actions. We use [ ] to denote intentions which, recall, are stacks of plans. Each plan in the stack will be separated by '|', and the leftmost plan is the top of the stack.

Assume that in one given reasoning cycle the selected event is $\langle +b, \top \rangle$ and the intended means is '+b : true <- !g; a_1.' which we call p_1. When the first formula of the plan body is executed (as we see in the later steps), because it is a goal to achieve, all the interpreter does is to create an event as follows:

$$\langle +!g, [+b \; : \; \texttt{true} \; \texttt{<-} \; !g; \; a_1.] \rangle$$

which says that the agent has come to have the new goal g because p_1 was intended and executed to a part of the plan body where that goal was required to be achieved. Assume now that in the reasoning cycle where the event above is the selected event, the intended means for that event turns out to be p_2 which is as follows: '+!g : true <- a_2.'. In this case, rather than creating a new intention as for external events, the intended means chosen for this internal event is *pushed on top of the intention that generated the event*, meaning that the following intention

---

[1]External events can also be goals delegated by another agent through communication, as we see in Chapter 6.

would be in the set of intentions at Step 8 of the reasoning cycle:

$$[\quad +!g \; : \; \texttt{true} \; \texttt{<-} \; a_2. \quad | \quad +b \; : \; \texttt{true} \; \texttt{<-} \; !g; \; a_1. \quad]$$

When this intention is executed, we will have the agent performing action a_2, which presumably gets goal g achieved, and only then can a_1 be executed. Recall also that goals (test goals especially) further instantiate variables in the plan body. So it should be now clear why an *intention* in AgentSpeak is defined as *a stack of partially instantiated plans*.

As a final comment, note that, whatever plan from the plan library is chosen to become the intended means, it is only an *instance* (i.e. a copy) of that plan that actually goes to the set of intentions. The plan library is not changed; it is the plan instance that is manipulated by the interpreter.

Step 9 – Selecting an Intention for Further Execution

Assuming we had an event to handle, so far in the reasoning cycle we acquired a new intended means. Further, as we saw above, typically an agent has more than one intention in the set of intentions, each representing a different focus of attention. These are all competing for the agent's attention, meaning that each one of them can potentially be further executed, which is what will be done in the next step of the reasoning cycle. However, in one reasoning cycle, at most one formula of one of the intentions is executed. Therefore, the next thing to do before the agent can act is to choose one particular intention among those currently ready for execution.

As might be expected from previous similar cases, again we shall use a selection function to make this choice; this selection function is called the *intention selection function* ($\mathcal{S}_\mathcal{I}$). Similarly to the other cases, the idea is that such decision is typically agent-specific, and details of customisation are given in Section 7.2. Any complex agent will require the choice of the particular intention to be executed next to be a well informed one. Typically, achieving certain goals will be more urgent than others, thus the choice of the next intention to get the agent's attention is really quite important for how effectively the agent will operate in its environment. In fact, one might want to use a sophisticated form of reasoning to make the best possible choice, for example following recent work on goal reasoning [96, 102] (we discuss this further in Chapter 11). However, all that *Jason* offers as pre-defined intention selection function is a form of 'round-robin' scheduling. That is, each of the intentions is selected in turn, and when selected, only *one* action (more precisely, one formula of the plan body) gets executed.

As for events, the set of intentions is implemented as a list, and to execute (one formula from) an intention, the pre-defined function removes the intention at the beginning of the list representing the set of intentions, and when inserted back (after the formula is executed at the next step of the reasoning cycle), it goes to the end of the list. This means that, unless the programmer customises this selection function, the agent will be dividing its attention equally among *all* of its intentions.

Step 10 – Executing One Step of an Intention

There are three main things that an agent does in every reasoning cycle: update its information about the world and other agents, handle one of the possibly many generated events and then, of course, act upon the environment (or, more generally, follow up on one of its intentions). It is this latter stage that is discussed here. We are going to (potentially) perform an action in the agent's environment. An agent typically has different intentions competing for its attention, but in the previous step we already selected the particular intention – which in the context here is one of the *courses of action* the agent committed to carrying out. In fact, an intention is a *partial* course of action because, remember, it involves having goals, but we only choose the courses of actions for those as late as possible, to wait until the agent has information about the world that is as updated as possible. Nevertheless, given one particular intention, it is fairly simple to decide what to do. It depends on the type of formula that appears in the beginning of the body of the plan that is at the top of the stack of plans forming that intention. Considering the example from Step 8, this will be much clearer. Suppose the intention selected in the previous step was:

$$[\quad \texttt{+!}g \; : \; \texttt{true} \; \texttt{<-} \; a_2. \quad | \quad \texttt{+}b \; : \; \texttt{true} \; \texttt{<-} \; \texttt{!}g; \; a_1. \quad]$$

which essentially means that the intention is formed by the plan we called p_2 on top of p_1. This is because p_1 cannot be continued before goal g is achieved, and p_2 is the plan the agent committed to execute (hence an intention) in order to achieve g. Thus p_2, the plan at the top of the intention, is the one that the intention determines that we can currently execute. Recall also that it is the *body* of the plan that gives us the particular course of action the agent has to execute. Therefore, the first formula in the body of plan p_2, namely 'a_2', is what the agent will now execute. In this example, the formula turns out to be an environment action, so the agent should then use its means of changing the environment (the 'effectors') in order to carry out the action represented by a_2. In most cases, after the formula is executed (or more generally 'dealt with'), it is removed from the body of the plan, and the (updated) intention is moved back to the set of intentions. As we see below, this is not the case for achievement goals, for example, but as a rule this is what happens, and that is why the formula at the *beginning* of the body of the topmost plan of the selected intention tells us what to execute next: the body is a *sequence* of formulæ to execute and when a formula is executed it is just deleted from the body of the plan.

It should now be clear why the execution of an intention depends on the type of formula we have in the beginning of the body of the plan at the top of the selected intention. Here is what the interpreter does depending on each type of formula. Recall that there are six different types of formulæ that can appear in a plan body:

> **Environment action.** An action in the plan body tells the agent that something should be effected in the environment. As we know, the reasoning process of the agents deals with the choice of the actions to take, but there is another part of the overall agent architecture which deals with the actual acting, called the agent's *effectors*. Recall also that we expect that part of the agent's architecture which deals with effecting changes in the environment to give the reasoning process (the AgentSpeak interpreter) a simple feedback on whether it did or did not execute the action. What *Jason* does, therefore, is to *suspend the intention* until the action execution is done, so that the agent can use this time to do other things. In practice, this means that the action formula is removed from the plan body and inserted in a special structure of the interpreter accessible by the method that interfaces with the effectors (the act method in Figure 4.1). Because we cannot immediately execute the next action of that plan — as we are waiting for the effector to perform the action and confirm to the reasoner whether it was executed or not — and the body of a plan needs to be executed in *sequence*, the intention needs to be suspended, so instead of returning it to the set of intentions, as is normally done, it goes to another structure that stores all intentions that were suspended because they are waiting for an *action feedback* or a *message feedback*: as we shall see in Chapter 6, certain types of communication require the agent to wait for a reply before that intention can be executed any further.
>
> One important thing to note is that, in this and some other cases, we are suspending *one particular intention*. An agent typically has various focuses of attention (separate intentions) and new events to handle, so even if some of the intentions are currently suspended, it might well be the case that in the very next reasoning cycle there will be some other intention to be further executed. This is to emphasise that an agent is unlikely to become idle because intentions are suspended in this way.
>
> **Achievement goals.** Having new goals to achieve creates what we have described as *internal events*. We saw earlier in Step 8 that internal events have the particular intention that generated the event as its second component. Therefore, when the interpreter executes an achievement goal formula, it generates the event as shown in Step 8

and the intention goes (as part of the event) to the set of events rather than returning to the set of intentions. Again, this is because the intention needs to be *suspended*: we cannot further execute it because we first need to find another plan to achieve that (sub)goal. Note that an intention suspended because an internal event is created goes to the set of events, not the set of suspended intentions (which is exclusively for intentions waiting action or communication feedback). When the literature on the original version of AgentSpeak uses the term 'suspended intentions', it is to this particular situation it is referring.

Note also that, unlike other formulæ, goals are not immediately removed from the body of the plan. They are not removed when the internal event is created but only when a plan to achieve the goal successfully finishes executing. Referring again to the example used in Step 8, note that the formula !g is still in the body of plan p_1 when p_2 is pushed on top of it. This is because, when the plan to achieve g is finished, we might need to use the unifying function of plan p_2 to instantiate any variables that were free (i.e. uninstantiated) in g when the event for +!g was generated. Only then can we remove the goal from the plan body, so that the intention is ready for the next formula to be executed. In the next section, we also discuss what happens when the plan to achieve a goal fails.

Test goals. In the previous chapter we saw that normally a test goal is used to check if a certain property is currently believed by the agent or to retrieve information that is already in the belief base. In these cases, if the test goal succeeds we can remove the goal from the plan body and the (updated) intention goes straight back into the set of intentions. Where the test goal is used to retrieve information from the belief base, this results in further instantiation of logical variables (as in the examples of checking beliefs in the plan context that we saw earlier). As this needs to be done by the interpreter itself (by trying to unify the goal with all the beliefs[2] in the belief base), this is done within the same reasoning cycle and therefore the intention could (if selected again) carry on executing in the next reasoning cycle.

However, recall that, if the required property/information is not present in the belief base, instead of failing the test goal the interpreter will try and see if there is a plan that is *relevant* for handling that test goal. For this, an internal event needs to be generated, exactly as

---

[2]*Jason* implements the belief base as a hash table indexed by the predicate symbol, which allows us to retrieve the likely candidates for unification very efficiently.

happens for achievement goals, explained above. The next section discusses what happens if that also fails or if an applicable plan is found but its execution fails.

Mental notes. If the formula is a belief to be added or deleted to/from the belief base, all the interpreter does is to pass the appropriate request for the brf method to do all necessary changes in the belief base and generate the respective events (belief additions and deletions) accordingly. Recall that the pre-defined brf method available with *Jason* carries out all such requests regardless of whether the belief base becomes inconsistent or not, and it does not do any other changes which would be required to maintain consistency; Section 11.2 discusses ongoing work on a more sophisticated alternative.

Normally, after sending the appropriate request to the brf method, the formula is removed from the plan body, and the intention is inserted back in the set of intentions (i.e. it can be selected for execution again in the next reasoning cycle). If the brf method actually makes any changes in the belief base, as when perception of the environment occurs, events are generated, which means they could trigger the execution of another plan. Whether that plan is executed as part of this intention or as a separate focus of attention can be configured by the programmer, as explained below in Section 4.4. Also, it is important to keep in mind that, *unless the programmer gives a specific* source *annotation* to the predicate in the belief addition/deletion formula, the interpreter will automatically add a source(self) annotation to any belief that is added/deleted to/from the belief base as a consequence of a 'mental note' type of formula in a plan body.

Internal actions. In this case, the Java/legacy code provided by the programmer is completely executed, the formula is removed from the body of the plan and the (updated) intention goes back to the set of intentions ready for execution in the next reasoning cycle. As for any other formula in the body, we need the notion of whether the 'action' succeeded or failed; this is why the code used to implement the internal action needs to be included within a *Boolean* Java method (the details are given in Chapter 7.1). As for goals, internal actions can also further instantiate variables that are still free in the plan body; this can be used if 'results' need to be returned by the internal action to be used in the agent's reasoning (the examples in Chapter 7 show the *Jason* support for instantiating logical variables in Java). Even though in principle any code can be used here, the only care programmers should have is that the whole code for the given internal action is executed within one reasoning cycle, so heavy computation (without another thread being created) may lead the agent not to react sufficiently fast to important events in the environment, which is likely to be inefficient/inappropriate for complex multi-agent systems. In Figure 4.1, we gave special attention to the .send internal action, which makes use of the sendMsg method provided by the communication infrastructure to send messages to other agents.

Advanced

The interpreter behaviour described above, of internal actions being removed from the plan body and the intention going back to the set of intentions ready for execution, is the normal behaviour for internal actions. However, some internal actions may require the interpreter to work differently, for example the .send predefined internal action used for communication. For certain types of communication requiring a reply, we may need to wait for it, so instead of the intention going back to the set of intentions as normal for internal actions, the intention goes to that special structure of suspended intentions, as for actions waiting for feedback from the effectors (the part of the overall agent architecture responsible for performing environment-changing actions).

Expressions. These are simply evaluated in the usual way. There is one thing to be careful about when using relational expressions in a plan body rather than in the context. While it is at times handy to use an expression (in particular with the = operator which attempts to unify the left- and right-hand side of the expression), if the expression evaluates to false the whole plan fails, and handling plan failure of course can be quite 'expensive'. On the other hand, if the expression denotes a property that is known to be required for the remainder of the plan execution to succeed anyway, and that can only be determined midway through the course of action, then of course including such an expression in the plan body might be useful.

Before another cycle starts, there are a few 'housekeeping' tasks that need to be done, as discussed next.

Final Stage before Restarting the Cycle

Some intentions might be in the set of suspended intentions either waiting for feedback on action execution or waiting for message replies from other agents. Before another reasoning cycle starts, the interpreter checks whether any such feedback and replies are now available, and if so, the relevant intentions are updated (e.g. further instantiation might occur) and included back in the set of intentions, so that they have a chance of being selected for further execution in the next reasoning cycle (at Step 9).

Another housekeeping task that might be necessary is to 'clear' intended means or intentions that might have executed to completion. If the formula executed in the current reasoning cycle was the last to appear in the body of the original plan in the plan library, that plan finished successfully so it must be removed from the top of the stack forming that intention. If that turned out to be the last plan in the intention, the whole intention is finished and can be removed from the set of intentions. If there is another plan below the finished plan in the stack, it is because that plan had an achievement or test goal which is now presumably achieved so that plan can now be resumed. However, before the finished plan is removed, the interpreter needs to check whether there were uninstantiated variables in the goal at the beginning of the body of the second plan in the stack. If so, then the triggering event of the finished plan is made ground (using the unifying function of the finished plan) and unified with the goal in the second plan in the stack. After this, the finished plan is removed from the stack and the goal at the beginning of the body of the current plan can also be removed (as it has just been achieved, presumably). Note that this removal means that we need to recursively attempt to clear the intention again. Only when the intention has been cleared to the point of being ready for a next formula to be executed can it be placed back in the set of intentions.

Advance

It is simple to observe that not all reasoning cycles will lead the agent to executing an action that changes the environment. Besides, for realistic applications, perceiving the environment and obtaining a symbolic representation of it can be a very computationally expensive task, which cannot be done as soon as the agent executes an action. Therefore, in many cases it might be interesting to configure the number of reasoning cycles the agent will perform before the **perceive** method is actually called to get updated sensorial information (unless configured otherwise, this is done again in every reasoning cycle). Section 4.3 shows that this can be done easily by setting an interpreter configuration option.

The agent is now ready to start another reasoning cycle (Step 1). We next discuss the plan failure handling mechanism, and after that we describe the mechanisms which allows programmers to configure certain aspects of the *Jason* interpreter.

4.2 Plan Failure

Multi-agent systems approaches for software development are aimed at application areas where the environment is dynamic and often unpredictable. In such environments, it will be common for a plan to fail to achieve the goal it was supposed to achieve. Therefore, in the same way that we give plans to agents so that they have alternative courses of action to choose from when they need to achieve a goal, we may need to give the agent plans for what to do when a preferred plan *fails*. The best way to think of this type of plan is as *contingency plans*. Whilst we use triggering events of the form +!g for plans to achieve g, we use the -!g triggering event notation to identify a contingency plan for when a plan for +!g fails.

Before we explain in detail how such contingency plans are written in AgentSpeak programs and how the interpreter handles plan failure, we need to understand the circumstances under which there can be a plan failure in the first place. Three main causes for plan failures can be identified:

Lack of relevant or applicable plans for an achievement goal. This can be understood as the agent 'not knowing *how* to achieve something desired (in its current situation)'. More specifically, this is the case where a plan is being executed which requires a subgoal to be achieved, and the agent cannot achieve that. This happens either because the agent simply does not have the know-how, where there are not even *relevant plans* for that subgoal – this could happen through a simple programming omission (the programmer did not provide the required plans) – or because all known ways of achieving the goal cannot currently be used (there are known plans but their contexts do not match the agent's current beliefs and so are not currently applicable).

Failure of a test goals. This represents a situation where the agent 'expected' to believe that a certain property of the world held, but in fact the property did not hold when required; these also represent conditions for the plan to continue successfully. If we fail to retrieve from the belief base the necessary information to satisfy a test goal, the interpreter further tries to generate an event for a plan to be executed, which then might be able to answer the test goal. If this also fails (for lack of relevant or applicable plans, as above), only then is the test goal deemed unsuccessful and therefore the plan where it appears fails.

Action failure. Recall that in *Jason* there are two types of actions: internal actions and basic (or environment) actions. The former are boolean Java methods (the boolean return value tells the interpreter whether the action was executed or not), whereas the latter represent the effectors within the agent architecture which are assumed to provide feedback to the interpreter (stating whether the requested action was successful or not). If an action fails, the plan where they appeared also fails.

Regardless of the reason for a plan failing, the *Jason* interpreter generates a goal deletion event – an event of the form '-!g' – if the plan for a corresponding goal achievement (+!g) has failed. We here give (informal) semantics to the notion of goal deletion as used in *Jason*. In the original AgentSpeak definition, Rao [80] syntactically defined the possibility of goal deletions as triggering events for plans (i.e. triggering events with -! and -? prefixes), but did not discuss what they meant. Neither was goal deletion discussed in further attempts to formalise AgentSpeak or its ancestor dMars [40, 41]. Our own choice was to use this as some kind of plan failure handling mechanism, as discussed below (even though this was probably not what it was originally intended for). We believe the 'goal deletion' notation (-!g) also makes sense for such plan failure mechanism; if a plan fails the agent would have to drop the goal altogether, so it is to try to prevent this from happening – or more specifically to handle the event of a goal having to be dropped – that plans of the form '-!g : ...' are written.

The intuitive idea is that a plan for a goal deletion is a 'clean-up' plan, executed prior to (possibly) 'backtracking' (i.e. attempting another plan to achieve the goal for which a plan failed). One of the things programmers might want to do within the goal deletion plan is to attempt *again* to achieve the goal for which the plan failed. In contrast to conventional logic programming languages, during the course of executing plans for subgoals, AgentSpeak programs generate a sequence of actions that the agent performs on the external environment so as to change it, the effects of which cannot be undone by simply backtracking: it may require further *action* in the environment in order to do so. Therefore, in certain circumstances, one would expect the agent to have to act so as to reverse the effects of certain actions taken before the plan failed, and only then attempt some alternative course of action to achieve that goal. This is precisely the practical use of plans with goal deletions (-!g) as triggering events.

It is important to observe that omitting possible goal deletion plans for a given goal addition implicitly denotes that such a goal should never be immediately backtracked – i.e. no alternative plan for it should be attempted if one fails. However, the failure 'propagates' to other plans within the intention so it is still possible that such a goal would be attempted later, possibly with a different plan – we explain this in more detail later in this section by means of an example. To specify that backtracking should *always* be attempted (for example, until a more specific plan become applicable which uses internal actions to explicitly cause the intention to be dropped), all the programmer has to do is to specify a goal deletion plan, for a given goal addition +!g, with empty context and the same goal in the body, as in the following example:

```
-!g : true <- !g.
```

When a failure happens, the whole intention is dropped if the triggering event of the plan being executed was neither an achievement nor a test goal *addition*: only these can be attempted to recover from failure using the goal deletion construct.[3] In cases other than goal aditions (+!g), a failed plan means that the whole intention cannot be achieved. If a plan for a goal addition fails, the intention *i* where that plan appears is suspended and goes to the set of events with the respective goal deletion event (⟨-!g, i⟩). Eventually, this might lead to the goal addition being attempted again as part of the plan to handle the -!g event. When a plan for -!g finishes, not only itself but also the failed +!g plan below it are removed from the intention (recall that an intention is a stack of plans). It is interesting to leave the failed plan in the intention, for example so that programmers can check which was the plan that failed (e.g. by means of *Jason* internal actions). As will be clear later, it is a programmer's decision to attempt the goal again or not, or even to drop the whole intention (possibly with special internal action constructs, whose informal semantics

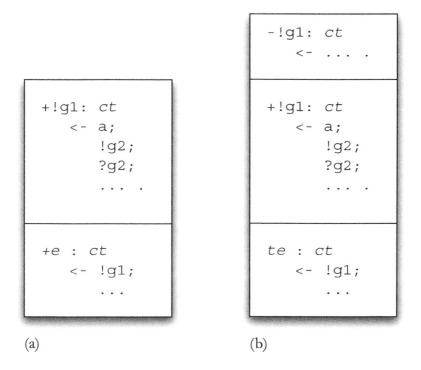

(a)   (b)

Figure 4.2 Plan failure due to an error in a. (a) An intention before plan failure; (b) that intention after plan failure.

---

[3] Note it is inappropriate to have a goal deletion event posted for a failure in a goal deletion plan, as this could easily cause a loop in executing the intention.

is given below), depending on the circumstances. What happens when a plan fails can be more clearly understood by the abstract example shown in Figure 4.2.

In the circumstance described in Figure 4.2(a), suppose a fails, or otherwise after that action succeeds an event for +!g2 was created but there was no applicable plan to handle the event, or ?g2 is not is the belief base, nor there are applicable plans to handle a +?g2 event. In any of those cases, the intention is suspended and an event for -!g1 is generated. Assuming the programmer included a plan for -!g1, and the plan is applicable at the time the event is selected, the intention will eventually look as in Figure 4.2(b). Otherwise, the failure propagates to the goals below g1, if there are any; this will be better explained by an example later in this section. If the failure propagates back to a plan for an *external event* (i.e. a plan with triggering event for belief addition or deletion) that event is re-posted or the whole intention dropped, depending on a setting of the *Jason* interpreter that is configurable by programmers.

Advance

Absence of relevant plans

One particular case of plan failure deserves more attention, namely that of lack of *relevant plans*. In the case of a test goal addition, there being no *relevant plans* means that the test goal was probably supposed to be a straight information retrieval from the belief base, and an internal event was generated just to check if the goal was not the trigger of a plan (i.e. a more complex query) and the fact that there are no relevant plans shows that this was not the case. For goal deletions (both achieve and test), a lack of relevant plans is used here to denote that the programmer did not wish for the corresponding goal addition to be immediately backtracked, i.e. to be attempted again. The case where there are no relevant plans for an external event is normally reasonable to expect: not every perceived change in the environment will lead the agent to new goals, and consequently intentions. In general, the lack of relevant plans for an event indicates that the perceived event is not significant for the agent in question, so they are simply ignored.

An alternative approach for handling the lack of relevant (and possibly applicable) plans is described in [6], where it is assumed that, in some cases explicitly specified by the programmer, the agent will want to ask other agents how to handle such events. The mechanism for plan exchange between AgentSpeak agents presented in that paper allows the programmer to specify which triggering events should generate attempts to retrieve external plans, which plans an agent agrees to share with others, what to do once the plan has been used for handling that particular event instance, and so on.

When an external event cannot be currently handled due to lack of *applicable plans*, one possibility is to place it back in the set of events, as later attempts at handling it may succeed. Note that this is different from there not being relevant plans for an event. The fact that the agent has (any) plans to handle an event means that this is likely to be an important task for the agent, but in the current state of affairs none of the known courses of action can be taken, so it may be worth trying again later on. This of course requires that the event selection function does not keep selecting this same event continuously – for the standard event selection function this is usually not a major problem as events are kept in a queue. Another obvious alternative is to disregard the event, i.e. just delete it. As we mentioned above, in *Jason* this can be configured by the user so that the interpreter uses either alternative by default, and this can also be configured for particular plans, using a standard plan annotation (as explained below in Section 4.3). See [5] for an overview of how various BDI systems deal with the problem of there being no applicable plans.

An Example of Plan Failure Handling

The use of plans for handling failure can be made easier to understand with an (abstract) example. The example shows how the internal state of the agent, especially the set of intentions, changes during its execution when some plan fails. The AgentSpeak code of this agent is written simply to exemplify such internal changes; the agent therefore has no complex plans, nor even interaction with the environment or other agents.

```
!g1. // initial goal

@p1 +!g1     : true <- !g2(X); .print("end g1 ",X).
@p2 +!g2(X) : true <- !g3(X); .print("end g2 ",X).
@p3 +!g3(X) : true <- !g4(X); .print("end g3 ",X).
@p4 +!g4(X) : true <- !g5(X); .print("end g4 ",X).
@p5 +!g5(X) : true <- .fail.

@f1 -!g3(failure) : true <- .print("in g3 failure").
```

The initial goal g1 creates a subgoal g2, which creates subgoal g3, and so on. The plan to achieve goal g5 uses the standard internal action .fail to force a failure in the achievement of g5. Among all these goals, only g3 has a plan to handle failure.

When the agent starts running, it has the initial goal !g1 added in the set of events. Plan @p1 is the only relevant and applicable plan for the +!g1 event (see Steps 6–8 of the reasoning cycle). A new intention is thus created based on plan @p1. The execution of this intention then generates the event +!g2(X) and plan @p2 is selected to handle it. The intention stack has now two intended means and looks as follows:

```
+!g2(X)
    <-  !g3(X); .print("end g2 ", X).
```

```
+!g1
    <-  !g2(X); .print("end g1 ", X).
```

The execution of the intention goes on until the activation of g5, by which time the intention looks like this:

```
+!g5(X)
    <-  .fail.
```

```
+!g4(X)
    <-  !g5(X); .print("end g4 ", X).
```

```
+!g3(X)
    <-  !g4(X); .print("end g3 ", X).
```

```
+!g2(X)
    <-  !g3(X); .print("end g2 ", X).
```

```
+!g1
    <-  !g2(X); .print("end g1 ", X).
```

As !g5(X) fails, the interpreter looks for a goal in the intention that has a relevant failure handling plan; in this example, only !g3(X) has such a plan. An event for -!g3(X) is then created and plan @f1 executed (note that variable X is bound to failure as -!g3(X) in the event was unified with -!g3(failure) in plan @f1).

```
-!g3(failure)                                        {X ↦ failure}
    <-  .print("in g3 failure").
```

```
+!g5(X)
    <-  .fail.
```

```
+!g4(X)
    <-  !g5(X); .print("end g4 ", X).
```

```
+!g3(X)
    <-  !g4(X); .print("end g3 ", X).
```

```
+!g2(X)
    <-  !g3(X); .print("end g2 ", X).
```

```
+!g1
    <-  !g2(X); .print("end g1 ", X).
```

When plan @f1 finishes, the intention continues in the same way it does when goal g3 is achieved successfully. That is, the second formula in the body of the plan for +!g2 is executed:

```
+!g2(X)                                    {X ↦ failure }
   <-  .print("end g2 ", X).
```
```
+!g1
   <-  !g2(X); .print("end g1 ", X).
```

The result of the above execution would be:

```
[a] saying: in g3 failure
[a] saying: end g2 failure
[a] saying: end g1 failure
```

We need to explain in more detail what happens when a plan fails, so we will refer again to the failure of the plan for +!g5 (plan @p5) described above. When that plan failed, an event ⟨-!g5, i⟩ (where i is the intention with @p5 at the top, as shown above) was generated. Because there are no *relevant* plans for -!g5, goal g5 has to be dropped with failure, which implies that plan @p4 cannot be further executed, so goal g4 has to be dropped, unless the programmer specified a contingency plan for g4 (which is not the case); if such a plan existed, it would match a -!g4 event. That is why the failure 'cascades' all the way down to g3, and the event effectively generated is -!g3.

When there are no *applicable* plans for a goal *deletion* or external events (i.e. belief additions or deletions), the interpreter can do one of two things: discard the whole intention or repost the event – by including it again at the end of the event queue – so that it can be tried again later. It is conjectured that each option will simplify programming (or will be more natural to use) in different applications. The way the interpreter works in such cases can be configured for each individual agent; configuration of the interpreter is the subject we discuss next.

4.3 Interpreter Configuration and Execution Modes

As we saw in earlier examples, *Jason* uses a project file where a multi-agent system is defined. The most general form of such definition is:

```
MAS <mas_name> {

    infrastructure: <Centralised|Saci|Jade|...>

    environment: <environment_simulation_class> at <host>
```

```
executionControl: <execution_control_class> at <host>

agents: <ag_type1_name> <source_file> <options>
                        agentArchClass <arch_class>
                        agentClass <ag_class>
                        beliefBaseClass <bb_class>
                        #<num_instances> at <host>;
        <ag_type2_name> ...;
}
```

Most parts of the multi-agent system definition file will be explained through-out the book, in the most appropriate chapters. In this section we concentrate on the options available for configuration of AgentSpeak interpreter that will be used in each agent, and also the overall *Jason* execution modes. The complete grammar for the language that can be used in defining multi-agent systems is given in Appendix A.2.

The interpreter configuration is done in the <options> part. The following settings are available for the AgentSpeak interpreter available in *Jason* (they are followed by '=' and then one of the associated keywords, where an underline denotes the option used by default):

events: options are discard, requeue or retrieve; the discard option means that external events for which there are no applicable plans are discarded (a warning is printed out to the console), whereas the requeue option is used when such events should be inserted back at the end of the list of events that the agent needs to handle. An option retrieve is also available; when this option is selected, the user-defined selectOption function is called even if the set of relevant/applicable plans is empty. This can be used, for example, for allowing agents to request plans from other agents who may have the necessary know-how that the agent currently lacks, as proposed in [6].

intBels: options are either sameFocus or newFocus; when internal beliefs (i.e. mental notes) are added or removed explicitly within the body of a plan, if the associated event is a triggering event for a plan, the intended means resulting from the applicable plan chosen for that event can be pushed on top of the intention (i.e. focus of attention) which generated the event, or it could be treated as an external event (as the addition or deletions of belief from perception of the environment), creating a new focus of attention. Because this was not considered in the original version of the language, and it seems to us that both options can be useful, depending on the domain area, we left this as an option for the user. For example, by using `newFocus` the user can create, as a consequence of a single external event, different intentions that will be competing for the agent's attention.

nrcbp: number of reasoning cycles before perception. Not all reasoning cycles will lead the agent to executing an action that changes the environment, and unless the environment is extremely dynamic, the environment might not have changed at all while the agent has done a single reasoning cycle. Besides, for realistic applications, perceiving the environment and obtaining a symbolic representation of it can be a very computationally expensive task. Therefore, in many cases, it might be interesting to configure the number of reasoning cycles the agent will have before the perceive method is actually called to get updated sensorial information. This can be done easily by setting the `nrcbp` interpreter configuration option. The default is 1, as in the original (abstract) conception of AgentSpeak(L) – that is, the environment is perceived at every single reasoning cycle.

The parameter could, if required, be given an artificially high number so as to prevent perception and belief update ever happening 'spontaneously'. If this is done, programmers need to code their plans to actively perceive the environment and do belief update, as happens in various other agent-oriented programming languages. See also page 254, where the internal action `.perceive()` is described.

verbose: a number between 0 and 2 should be specified. The higher the number, the more information about that agent (or agents if the number of instances is greater than 1) is printed out in the *Jason* console. The default is in fact 1, not 0; verbose 1 prints out only the actions that agents perform in the environment and the messages exchanged between them. Verbose 0 prints out only messages from the .print and .println internal actions. Verbose 2 is the 'debugging' mode, so the messages are very detailed.

user settings: Users can create their own settings in the agent declaration, for example:

```
... agents: ag1 [verbose=2,file="an.xml",value=45];
```

These extra parameters are stored in the Settings class and can be checked within the programmer's classes with the getUserParameter method, for example:

```
getSettings().getUserParameter("file");
```

The use of this type of configuration is rather advanced and its useful-
ness will be clearer in later chapters of the book.

The other configuration topic we discuss here is in regards to *execution
modes* of the *Jason* platform, which can be determined, if necessary, following
the `executionControl` keyword. Currently, there are three different execution
modes available in the *Jason* distribution:

- **Asynchronous**: all agents run asynchronously. An agent goes to its next
  reasoning cycle as soon as it has finished its current cycle. This is the *default*
  execution mode.

- **Synchronous**: all agents perform one reasoning cycle at every 'global execution step'. When an agent finishes its reasoning cycle, it informs *Jason*'s
  controller and waits for a 'carry on' signal. The *Jason* controller waits until
  all agents have finished their reasoning cycles and then sends the 'carry on'
  signal to them.

  To use this execution mode, the user has to set up a controller class in the
  multi-agent project/configuration file, as follows:

```
MAS test {
      infrastructure: Centralised
      environment: testEnv

      executionControl: jason.control.ExecutionControl

      agents:  ...
}
```

  The jason.control.ExecutionControl class implements precisely the Synchronous
  '*Jason* controller' described above.

- **Debugging**: this execution mode is similar to the synchronous mode, except
  that the *Jason* controller will also wait until the user clicks the 'Step' button
  before sending the 'carry on' signal to the agents.

  To use this execution mode you can just press the 'debug' button rather than
  the 'run' button of the *Jason* IDE, or you can to set up a debug controller
  class in the .mas2j configuration, for example:

```
MAS test {
      infrastructure: Centralised
      environment: testEnv
```

```
        executionControl: jason.control.ExecutionControlGUI


        agents:  ...
    }
```

The jason.control.ExecutionControlGUI class implements the *Jason* controller that creates a GUI for debugging. The graphical tool that helps debugging is called *Jason*'s 'Mind Inspector', as it allows users to observe all changes in agents' mental attitudes after a reasoning cycle. This also applies to distributed agents (e.g. using SACI, or JADE [9]).

Besides, the three available execution modes, it is possible for users to develop their own execution modes. Users with specific requirements in controlling agents' execution can define an ExecutionControl sub-class and specify it in project configuration file.

You will most likely have to override the following methods:

```
public void receiveFinishedCycle(String agName, boolean breakpoint) {
    super.receiveFinishedCycle(agName, breakpoint);
    ... your code ...
}
protected void allAgsFinished() {
    ... your code ...
}
```

These methods are called by *Jason* when one agent has finished its reasoning cycle and when all agents have finished the current 'global execution step' respectively.

To signal the agents to 'carry on', the user's class can use the following code:

```
    infraControl.informAllAgsToPerformCycle();
```

It is advisable that programmers who need to do this use *Jason*'s ExecutionControlGUI class as an example, and also read the API documentation for further available methods inherited from the ExecutionControl class.

4.4 Pre-Defined Plan Annotations

As mentioned in Chapter 3, annotations in plan labels can be used to associate meta-level information about the plans, so that the user can write selection functions that access such information for choosing among various relevant/applicable plans or indeed intentions. Further to that, *Jason* provides four pre-defined annotations which, when placed in the annotations of a plan's label, affect the way that plan is interpreted. These pre-defined plan label annotations are as follows:

atomic: if an instance of a plan with an `atomic` annotation is chosen for execution by the intention selection function, this intention will be selected for execution in the subsequent reasoning cycles *until that plan is finished* – note that this overrides any decisions of the intention selection function in subsequent cycles (in fact, the interpreter does not even call the intention selection function after an `atomic` plan has begun to be executed). This is useful in cases where the programmer needs to guarantee that no other intention of that agent will be executed in between the execution of the formulæ in the body of that plan.

breakpoint: this is useful in debugging: if the `debug` mode is being used (see Section 4.3), as soon as any of the agents start executing an intention with an instance of a plan that has a `breakpoint` annotation, the execution stops and the control goes back to the user, who can then run the system step-by-step or resume normal execution.

all_unifs: is used to include all possible unifications that make a plan applicable in the set of applicable plans. Normally, for one given plan, only the first unification found is included in the set of applicable plans. In normal circumstances, the applicable-plan selection function is used to choose between *different* plans, all of which are applicable (as we see in the next chapter). If you have created a plan for which you want the applicable-plan selection function to consider which is the best unification to be used as intended means for the given event, then you should included this special annotation in the plan label.

priority: is a term (of arity 1) which can be used by the default plan selection function to give a simply priority order for applicable plans or intentions. The higher the integer number given as parameter, the higher the plan's priority in case of multiple applicable plans. Recall you can implement more sophisticated selection functions and still make use of this (and other user-defined) annotations if you want. (Note that this is as yet unavailable, but a reserved plan annotation.)

4.5 Exercises

1. Suppose that an agent has the following plan library: | Basic |

```
@p1   +g(X,Y) : true <-  ...  .
@p2   +g(X,Y) : a(Y) & not b(X) <-  ...  .
@p3   +g(X,_) : a(Y) & Y > X <-  ...  .
```

```
@p4   +g(X,Y)[source(self)] : true <-  ...  .
@p5   +g(X,Y)[source(self),source(ag1)] : true <-  ...  .
@p6[all_unifs]   +g(10,Y) : a(Y) <-  ...  .
```

and the following beliefs,

```
a(10).
a(5).
b(20).
```

If the event ⟨+g(10,5)[source(ag1)]⟩ is selected, which plans are relevant and which are applicable?

Basic

2. Considering that the state of the stack of an agent's intention is as shown below:

+!g2	: true <-	a3; a4.
+!g1	: true <-	!g2.
+b	: true <-	!g1; a1.

(a) What will be the state of the intention after the execution of action a3? And after a4?

(b) If the agent's plan library includes the following plan:

```
-!g1 : true <- a5.
```

what will be the state of the above intention if action a3 fails?

Basic

3. In the Domestic Robot example, described in Section 3.4, the robot has three plans to achieve the goal !has(owner,beer) (identified by labels @h1, @h2 and @h3). If any of those plans fail, or none is applicable, the robot will not achieve the goal and the owner will remain waiting for the beer forever. To avoid this problem, add a failure handling plan for the goal !has(owner,beer) that sends a message to the owner informing her/him about the robot being unable to achieve the goal.

Advanced

4. Consider a multi-agent system composed of two agents. One agent, called 'sender', sends 100 messages to another, called 'rec' (the receiver). The rec agent simply sums the number of received messages and, after receiving the last message, prints out how many messages it received.

The sender AgentSpeak code is

```
!send(100).                   // initial goal
+!send(0) : true <- true.     // stop sending
+!send(X) : true
   <- .send(rec,tell,vl(X)); // send value to rec
      !send(X-1).            // send next value
```

and the rec code is

```
sum(0).
+vl(1) : sum(S) <- .print(S+1). // the last tell from sender
+vl(X) : sum(S) <- NS = S + 1; -+sum(NS).
```

When we execute these two agents, the rec output is usually no less than 10, but it is never 100, the expected value. Try to fix this bug in the rec agent.

5

Environments

One of the key aspects of autonomous agents is that they are *situated* in an environment. In multi-agent systems, the environment is shared by multiple agents, so an agent's actions are likely to interfere with those of other agents. Therefore, having an explicit notion of *environment*, although not mandatory, is likely to be an important aspect in developing a multi-agent system. In this chapter we discuss the support that *Jason* gives for implementing an environment model in Java.

In many multi-agent applications, the environment is the real world, and the Internet is another typical 'environment' where agents are situated. As an example of a real-world environment, consider a multi-agent systems developed to control a manufacturing plant. When an agent decides how to act, the action itself will be performed by some hardware that effects changes in the manufacturing machinery. Equally, perception of the environment is obtained from sensors that capture aspects of actual properties of the current state of the manufacturing process.

However, in other cases, we need to create a computational model of a real-world or artificial environment, and we need to be able to simulate the dynamic aspects of that environment. We then implement the agents that will operate in this simulated environment, acting upon it and perceiving properties of the simulated environment. This is particularly important, for example, for *social simulation* [50, 51, 79, 93], an area of research that makes extensive use of multi-agent systems techniques. More generally, for any (distributed) system that is completely computational (say, the bookshop management system in Chapter 9, taken from [75]), it is often a good idea to make sure that the system design explicitly incorporates a notion of environment. This chapter will discuss how to implement such type of environments in particular.

Before we start discussing the details of how simulated environments can be programmed in Java to work with AgentSpeak agents in *Jason*, we feel it is very important to emphasise how useful simulated environments are even for applications that are aimed at deployment in real-world environments. As multi-agent systems techniques are normally used to develop complex distributed systems, verifying and validating such systems is a very hard task. In Section 11.2, we shall mention briefly ongoing work on formal verification (using model checking) for AgentSpeak multi-agent systems. However, validation techniques such as simulation and testing are more likely to be used for a variety of agent applications. Therefore, it is likely that most developers will want to have a Java model of the real-world environment with which they will be able, by simulation, to evaluate how well the system can perform under specific circumstances of the target environment. This is common practice for industrial applications of multi-agent systems; see, for example, the use of simulation to validate the industrial application described in [67].

Another important feature of *Jason* regarding deployment of the actual system after simulation with a model of the environment is the architecture customisation that will be discussed in Section 7.3. With this, deployment can be done by simply switching the implementation of the methods that interface the agent with the environment. Instead of acting upon/perceiving the simulated environment, one can simply change to another architecture that uses specific Java methods that allow the agent to act upon/perceive the target environment (assuming such implementation is available, of course).

Readers not very fond of Java may be wondering why we are implementing environments in Java rather than, for example, creating an agent in AgentSpeak to simulate the environment. The reason is that the abstraction of agent-oriented programming is excellent for implementing *agents*, that is, autonomous entities acting in an environment, but are *not* ideal for environments. It is likely that computational resources would be wasted by trying to use a sophisticated reasoning architecture to model an environment. Besides, Java will provide all the abstractions needed for environment models (e.g. objects) as well as having well known support for things such as graphical user interfaces, which are often useful for visualising an environment. On the other hand, there is ongoing work on developing a high-level language that is specifically designed for modelling and implementing environments for multi-agent simulations (see Section 11.2).

5.1 Support for Defining Simulated Environments

As discussed above, in most applications a (simulated) environment written in Java will be necessary. The multi-agent system is then built by putting a number of AgentSpeak agents to operate in the shared environment. We here discuss the support provided by *Jason* for the development of such environment models.

Each individual agent in a multi-agent system will have to interact, possibly directly with the other agents through speech-act based communication (see Chapter 6), but also with the environment. The overall architecture of an agent (which can be customised as explained in Section 7.3) will include Java methods which implement the interaction of an agent with its environment. This is clearly shown in the UML sequence diagram shown in Figure 5.1. Note that the default implementation of the architectures already provides the interaction with environments simulated in Java, as depicted in the UML sequence diagram. The interface between agent architecture and environment only needs to be changed if the environment is not implemented as a Java program.

The UML sequence diagram in Figure 5.1 shows that the existing agent architecture uses the getPercepts method to retrieve, from the simulated environment, the *percepts* to which that particular agent currently has access (i.e. the properties of the environment currently perceptible to that particular agent). This is done whenever the current step of the reasoning cycle is the one in which perception is to be done, and recall that, after that, belief update takes place. In Chapter 4, we explained that, when an intention is executed and the formula being executed is an environment action, the reasoner 'requests' the action execution and suspends the intention until there is a feedback from the environment saying whether the action

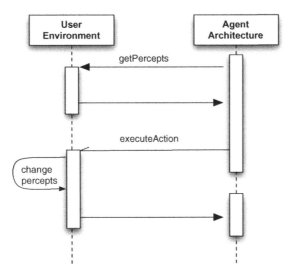

Figure 5.1 Interaction between an environment implementation and agent architecture.

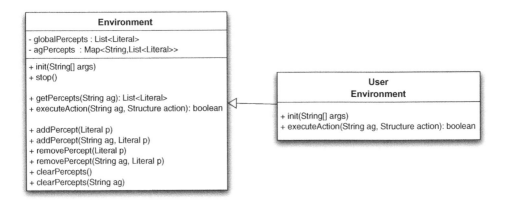

Figure 5.2 Implementation of an environment by a user application.

was executed or not.[1] For each such action execution request, the agent architecture invokes the executeAction method of the environment, and resumes the respective intention when the method returns; this is done while the reasoning cycle carries on.

Advance

Because the intention executing the action is suspended, and the reasoning cycle goes ahead while the architecture interacts with the environment, the effect is as if the executeAction method was invoked asynchronously. Note that the environment can be running in a different machine than the agent, therefore the suspended time of an intention can be significant in such cases.

The Environment class provided by *Jason* supports individualised perception, facilitating the task of associating certain percepts to certain agents only. The class keeps data structures used to store the percepts to which each agent has access as well as 'global' percepts that all agents can perceive. The method getPercepts checks the relevant data structures so that all percepts available to the particular agent calling the method are retrieved.

In order to implement an environment in *Jason*, programmers need to extend the Environment class and are likely to need to override the executeAction and init methods (Figure 5.2 depicts a class diagram relating these classes). A user-defined class implementing an environment typically looks as follows:

---

[1]This does not mean that the expected effects in the environment took place, but at least we know the action was attempted.

```
import jason.asSyntax.*;
import jason.environment.*;

public class <EnvironmentName> extends Environment {

    // any class members needed...

    @Override
    public void init(String[] args) {
        // setting initial (global) percepts ...
            addPercept(Literal.parseLiteral("p(a)"));

        // if open-world is begin used, there can be
        // negated literals such as ...
            addPercept(Literal.parseLiteral("~q(b)"));

        // if this is to be perceived only by agent ag1
            addPercept("ag1", Literal.parseLiteral("p(a)"));
    }

    @Override
    public void stop() {
        // anything else to be done by the environment when
        // the system is stopped...
    }

    @Override
    public boolean executeAction(String ag, Structure act) {
        // this is the most important method, where the
        // effects of agent actions on perceptible properties
        // of the environment is defined
    }
}
```

where <EnvironmentName> is the name of the environment class to be specified
in the multi-agent system configuration file.

The init method can be used to receive parameters for the environment class
from the multi-agent system configuration. It is also normally appropriate to use
this method to initialise the lists of percepts with the properties of the environment
that will be perceptible when the system starts running (i.e. the very first environment properties the agents will perceive when they start running). Percepts are
typically created using the parseLiteral method of the Literal class. Both positive and
negative percepts (for creating an open-world multi-agent system) can appear in a
list of percepts. Negative percepts are *literals* with the strong negation operator '~'.

These are the Java methods that can be used to program a *Jason* environment:

addPercept(L): add literal L to the global list of percepts; that is, *all agents* will perceive L if they sense the environment before L is removed from the global percepts;

addPercept(A,L): add literal L to the list of percepts that are exclusive to agent A; that is, only A will be able to perceive L;

removePercept(L): remove literal L from the global list of percepts;

removePercept(A,L): remove literal L from the list of percepts that are exclusive to agent A;

clearPercepts(): delete all percepts in the global list of percepts;

clearPercepts(A): delete all percepts in the list of percepts that are exclusive to agent A.

Only instances of the class Literal, which is part of the jason package, should be added to the lists of percepts maintained by the Environment class using these methods. Normally, one should *not* add any annotations here, as all percepts will be received by the agents with a source(percept) annotation automatically included.

Advance

> The access to the lists of percepts in the Environment class is automatically synchronised by the methods above (such as addPercept()). This is important given that agents will be concurrently executing the executeAction method and the percept structures are global for the Environment class. Although the basic work of synchronisation is done by *Jason*, depending on how the model of the environment is done, programmers may need further synchronisation in the executeAction method, which of course should be done carefully for the sake of the efficiency of its concurrent execution.

Most of the code for building environments should be (referenced at) the body of the method executeAction, which must be declared exactly as described above. Whenever an agent tries to execute an environment action, the name of the agent and a Structure representing the requested action are sent as parameters to this method. The code in the executeAction method typically checks the Structure (which has the form of a Prolog structure) representing the action (and any required parameters) being executed, and checks which is the agent attempting to execute the action. Then, for each relevant action/agent combination, the code does whatever is necessary in that particular model of an environment to execute the particular action by that particular agent. Normally, this means changing the percepts – i.e. what will be perceived as true or false in the environment – according to the action being performed (the examples given below show the typical form of such method implementation). Note that the execution of an action needs to return a boolean value, stating whether the agent's request for an action execution was attempted or refused by the environment. A plan fails if any basic action attempted by the agent fails to execute.

vanced

> It is worth making a note regarding environments with individualised perception – i.e. the fact that in programming the environment one can determine what subset of the environment properties will be perceptible to individual agents. Recall that within and agent's overall architecture further customisation of what beliefs the agent will actually acquire from what it perceives can be done. Intuitively, the environment properties available to an agent, determined by the environment definition itself, should be associated with what is actually perceptible in the environment for a particular agent; for example, if something is behind my office wall, I cannot see it. The customisation at the agent overall architecture level should be used for simulating faulty perception – even though something is perceptible for that agent in that environment, some of those properties may still not be included in the agent's belief update process, hence the agent will not be aware of them.

There are two typical errors made by beginners when programming multi-agent systems which are related to the model of the environment using the pre-defined *Jason* methods and classes. The first is that some programmers expect the agent to keep the perceived information in mind even if the percept only lasted for one reasoning cycle (and then was removed from the percepts in the environment model). If an agent needs to remember things after they are no longer perceived, plans have to be added to the agent that create *mental notes* as soon as the relevant property is perceived. Mental notes are 'remembered' until explicitly deleted. Perceived information is removed from the belief base as soon as they are no longer perceived as true of the environment. Also note that, if a percept is added in the environment model and then removed by another action, it is possible that some agents will never perceive that property, because they were in the middle of a single reasoning cycle when the percept was added and removed. On the other hand, if a property is perceived, recall that events are created by the agent architecture (the belief update process more specifically) both when the property is first perceived and as soon as it is no longer perceived.

The second typical error in programming agents and environments is mismatches (essentially typos) between percepts and action names as used in the environment model and as used in the agent AgentSpeak code. Programmers familiar with ontology technologies [91] who are not using an agent-oriented software methodology to guide the implementation may benefit from using an ontology of all the terms used in the agents, the environment, and for agent communication in order to avoid such errors.

Let us now consider a more concrete example of an environment. In Section 5.2 below, we give a more elaborate example. The one we use here is just to make clear the use of various methods available in the *Jason* Environment class. The example shows how to implement a *Jason* environment for the simplified 'vacuum cleaner world' used as an example in [82, p. 58]. In this very simple scenario, there are only two positions where the vacuum cleaner can be, identified as left and right. The vacuum cleaner can move to the left, to the right, and suck any dirt in either position. Therefore, the actions available to the vacuum cleaner are, respectively, left, right and suck. The only percepts that the agent controlling the vacuum cleaner will require for its reasoning are: (i) whether the current position is dirty; and (ii) the actual position itself (pos(l) for the left and pos(r) for the right position). Note also that there is no uncertainty or non-determinism in this very simple environment: dirty is perceived if and only if the position is dirty, the current position is also accurately perceived and all actions always succeed when they are valid (for example, moving to the right whilst at the right position obviously has no effect in this simple environment).

```
import jason.asSyntax.*;
import jason.environment.*;
import java.util.logging.*;
import java.util.*;

/** Simple Vacuum cleaning environment */
public class VCWorld extends Environment {

  // dirty[0] is the left loc, dirty[1] the right
  boolean[] dirty = { true, true };

  // the vacuum cleaner is in the left
  char vcPos = 0;

  Random r = new Random();
  Logger logger = Logger.getLogger("env.VC");

  @Override
  public void init(String[] args) {
     updatePercepts();
  }
  /** update the agent's percepts based on the current
      state of the world model */
```

```
private void updatePercepts() {
  // dynamically add dirty
  if (r.nextInt(100) < 20) {
    dirty[r.nextInt(2)] = true;
  }
  clearPercepts(); //remove previous percepts
  if (dirty[vcPos]) { //'dirty' must be added before position as
                     // the agent model assumes this order is used
    addPercept(Literal.parseLiteral("dirty"));
  } else {
    addPercept(Literal.parseLiteral("clean"));
  }
  if (vcPos == 0){
    addPercept(Literal.parseLiteral("pos(l)"));
  } else if (vcPos == 1) {
    addPercept(Literal.parseLiteral("pos(r)"));
  }
}
```

Some of the Java methods used in the example above are the ones most likely to be used for implementing environments. In particular, note how addPercept combined with Literal.parseLiteral is used to create percepts, clearPercepts is used to delete all previous (global) percepts, and getFunctor is used to return (as a string) the atom that is the head (i.e. the *functor*) of the structure representing the action. In this example the actions are so simple that we only have an atom, without any parameters (i.e. a structure of arity 0) to represent each of the actions. *Jason* uses the Java logging API to print (and store) messages, so instead of System.out.println it is more appropriate to use logger.info. The logging API allows the programmer to customise the format, level and device to which the messages will be printed. The next section gives a more elaborate example of an environment for *Jason* agents.

5.2 Example: Running a System of Multiple Situated Agents

The environment of the vacuum cleaner example is very simple, so we can implement it all with a single Java class. However, most of the scenarios of interest for multi-agent systems have environments that are more complex than that, and therefore require a more careful environment design. In this section, we describe how the environment of the example in Section 3.4, which is not as simple as the vacuum cleaner, can be designed and implemented. Its design is based on a common object-oriented design pattern called *Model – View – Control*, normally used for graphical interfaces but also suitable for the design of *Jason* environments. The environment design is thus based on the following three components:

1. The *model* maintains the information about the environment state (e.g. a robot's location) and the dynamics of the environment (e.g. the change in the robot's location when it moves).

2. The *view* renders the model into a form suitable for visualisation.

3. The *controller* interacts with the agents and invokes changes in the model and perhaps the view.

Figure 5.3 shows the class diagram of an environment design for the domestic robot scenario. To provide the percepts as shown in Figure 3.3, the model of this example environment should maintain: (i) the number of available beers in the fridge; (ii) whether the owner currently has a bottle of beer; and (iii) the robot's location. The number of beers is a simple attribute of the model, (availableBeers in Figure 5.3). The percept has(owner,beer) is based on the sipCount value. When the owner gets a beer, sipCount is initialised with 10 and this value decreases by 1 each time the sip(beer) action is executed. That percept is thus available to both the robot and the owner while sipCount > 0. Although we only need to provide for perception of the robot's location as being either at the fridge or next to the owner, we will use the GridWorldModel class, provided by the *Jason* API, to maintain the robot's location. This class represents the environment model as an $n \times m$ grid where each position may contain objects. The main advantage of using this model is that the 'view' is given (see Figure 5.4 for the view of the domestic robot

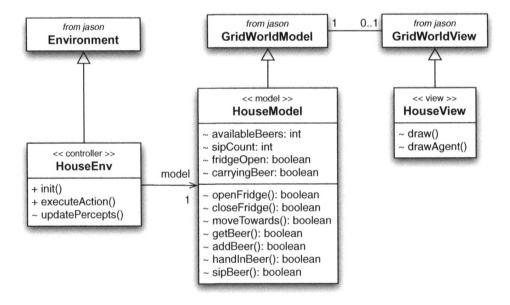

Figure 5.3 Class diagram of the domestic robot environment.

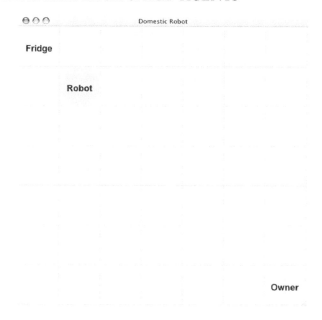

Figure 5.4 Graphical user interface of the domestic robot environment.

application), so we do not need to implement most of the graphical interface. The GridWorldModel class implements two methods: one to draw the static objects (such as the fridge in the domestic robot scenario) and another to draw the (visible aspects of) agents.

The model also has two other boolean attributes: fridgeOpen and carryingBeer. The former represents the state of the fridge door and is used to decide whether to give the stock(beer,N) perception to the robot or not, since only when the fridge is open is the stock perceived. The carryingBeer attribute is used to represent whether the robot is carrying a beer; the action hand_in(beer) should only succeed if the robot is carrying one.

The model implements one method for each possible action in the environment. Such methods should check the action pre-conditions, change the attributes accordingly, and return a boolean value stating whether the action was executed or not. As an example, we give an excerpt of the Java code of some of these methods:[2]

```
boolean getBeer() {
    if (fridgeOpen && availableBeers > 0 && !carryingBeer) {
        availableBeers--;
        carryingBeer = true;
        return true;
    } else {
```

---

[2]The complete code of the class is available on the web page for this book.

```
          return false;
       }
    }


    boolean handInBeer() {
       if (carryingBeer) {
          sipCount = 10;
          carryingBeer = false;
          return true;
       } else {
          return false;
       }
    }
```

The controller in this example should provide the percepts to the agents and receive their actions. The easiest way to implement the controller in *Jason* is to extend the Environment class. This class can then create the model and eventually the view components, for example using the init method:

```
public void init(String[] args) {
   model = new HouseModel();

   if (args.length == 1 && args[0].equals("gui")) {
      HouseView view  = new HouseView(model);
      model.setView(view);
   }

   updatePercepts();
}
```

This method first creates the model and, if the user has used 'gui' in the system project configuration, it also creates the view, so that the user can easily turn the graphical interface on and off. Note that the system project for the domestic robot example has an environment keyword followed by the class name that implements the environment and its parameters, as follows:

```
MAS domestic_robot {

    environment: HouseEnv(gui)

    agents: robot; owner; supermarket;
}
```

The updatePercepts() method maps the model information into percept formulæ. In the code below, the method model.getAgPos (from GridWorldModel) returns the location of the robot (represented in the model as agent index 0);

model.lFridge represents the fridge location (an $\langle x, y \rangle$ coordinate); model.lOwner represents the owner location; `af` is the literal `at(robot,fridge)`; `ao` is the literal `at(robot,owner)`; and `hob` is `has(owner,beer)`.

```
void updatePercepts() {
  // clear the perceptps of the agents
  clearPercepts("robot");
  clearPercepts("owner");

  // get the robot location
  Location lRobot = model.getAgPos(0);

  // add agent location to its percepts
  if (lRobot.equals(model.lFridge)) {
     addPercept("robot", af);
  }
  if(lRobot.equals(model.lOwner)) {
     addPercept("robot", ao);
  }
  // add beer "status" to percepts
  if (model.fridgeOpen) {
     addPercept("robot",
        Literal.parseLiteral(
           "stock(beer,"+model.availableBeers+")"));
  }
  if (model.sipCount > 0) {
     addPercept("robot", hob);
     addPercept("owner", hob);
  }
}
```

Finally, the executeAction method simply calls the methods in the model:

```
public boolean executeAction(String ag, Structure action) {
   boolean result = false;
   if (action.equals(of)) {  // of = open(fridge)
      result = model.openFridge();

   } else if (action.equals(clf)) { // clf = close(fridge)
      result = model.closeFridge();

   } else if (action.getFunctor().equals("move_towards")) {
      String l = action.getTerm(0).toString();
      Location dest = null;
      if (l.equals("fridge")) {
         dest = model.lFridge;
      } else if (l.equals("owner")) {
```

```
        dest = model.lOwner;
    }
    result = model.moveTowards(dest);
  ...
  }

  if (result)
    updatePercepts();

  return result;
}
```

5.3   Exercises

1. Considering the vacuum cleaner scenario used earlier:

   (a) Write the AgentSpeak code for the agent controlling the vacuum | Basic |
       cleaner.

   (b) Change the given environment to add perception only for the robot, | Basic |
       rather than all agents; i.e. change it from global to 'individualised' per-
       ception.

   (c) Increase the environment model to a 10 × 10 grid using the same design | Advanc |
       pattern of the domestic robot system.

   (d) Implement the AgentSpeak code for the agent in the 10 × 10 envi- | Advanc |
       ronment and try more instances of the vacuum cleaner agents in the
       system.

2. Considering the domestic robot scenario:

   (a) In the implementation of the domestic robot environment shown in | Basic |
       this chapter, we did not check who was the agent requesting the action.
       For instance, if the robot chooses the action sip(beer), the environ-
       ment will execute the action as if it had been requested by the owner.
       Change the environment implementation so that only the right agent
       is allowed to perform each action (as specified in Figure 3.3).

   (b) Remove the move_towards(Place) action and add four new actions | Basic |
       for the robot: up, down, left, and right. Note that in this case
       the location perception for the robot also needs to be improved (e.g.
       at(robot,Column,Line)) so that it can decide what is the best action
       to take.

vanced (c) Add obstacles to the environment of the dometic robot so that the robot can perceive those obstacles that are near enough. The choice of actions up, down, left and right should be constrained by the obstacles.

vanced (d) Change the domestic robot environment so that it becomes non-deterministic. Some actions, deliver(beer,N) for instance, will be correctly performed only, say, 95% of the time. Also update the supermarket code to deal with this uncertainty. Note that the supermarket will need some perceptual feedback from the environment, delivered(beer,N) for instance, in order to know whether its action was performed or not. Another solution is for the supermarket to wait for a confirmation message from the robot, but we will leave this for the next chapter.

vanced 3. Suppose an MAS environment is a room with a single door. Inside this room we have three agents. One agent is claustrophobic and insists on keeping the door unlocked. The others are paranoid and want to keep the door locked.

This environment implementation is simply:

```
import jason.asSyntax.*;

public class RoomEnv extends jason.environment.Environment {

    Literal ld  = Literal.parseLiteral("locked(door)");

    public void init(String[] args) {
        addPercept(ld); // initial perception: door locked
    }

    public boolean executeAction(String ag, Structure act) {
        if (act.getFunctor().equals("unlock"))
            clearPercepts();
        if (act.getFunctor().equals("lock"))
            addPercept(ld);
        return true;
    }
}
```

The agents' code is even simpler – the claustrophobe's code is:

```
+locked(door) : true <- unlock.
```

and the code for the paranoids is:

```
-locked(door) : true <- lock.
```

The project file is

```
MAS room {
    environment: RoomEnv
    agents: claustrophobe;
            paranoid #2;
}
```

The '#2' after the paranoid states that two instances of this agent will be created; the first agent is named 'paranoid1' and the second 'paranoid2'.

When this system is executed, the result could be as follows:

```
Running project room
[claustrophobe] doing: unlock
[paranoid1]     doing: lock
[claustrophobe] doing: unlock
[paranoid1]     doing: lock
[claustrophobe] doing: unlock
[paranoid1]     doing: lock

  . . .
```

Why does only paranoid1 react to the deletion of the percept locked (door) caused by the claustrophobe's action?

6

Communication and Interaction

In Chapter 2, we saw that communication in multi-agent systems is typically based on the *speech act* paradigm, and we briefly reviewed the KQML and FIPA agent communication languages. In this chapter, we see how agent communication is accomplished in *Jason*.

Recall that in Section 4.1 we saw that, at the beginning of each reasoning cycle, agents check for messages they might have received from other agents. Any message received[1] by the checkMail method (see Figure 4.1) has the following structure:

$$\langle sender, illoc\_force, content \rangle$$

where *sender* is the AgentSpeak atom (i.e. a simple term) which is the name by which the agent who sent the message is identified in the multi-agent system; *illoc_force* is the *illocutionary force*, often called *performative*, which denotes the intention of the sender; and *content* is an AgentSpeak term which varies according to the illocutionary force. In this chapter, we see each of the available performatives and how they are interpreted by *Jason*.

While *Jason* interprets automatically any received messages according to the formal semantics given in [100], it is the programmer's responsibility to write plans which make the agent take appropriate communicative action. Ideally, future work on agent communication and agent programming languages should lead to the development of tools that allow high-level specifications of agent interactions (i.e. protocols) to be automatically transformed into plan skeletons which ensure the agent is capable of following the relevant protocols.

---

[1] Note that the format of the message from the point of view of the sending and receiving agents is slightly different; for example, a message to be sent has a reference to the receiver whereas a received message has a reference to the sender. The format explained here is that from the point of view of the *receiver* specifically.

6.1   Available Performatives

Messages are sent, as we saw in Chapter 3, with the use (in plan bodies) of a special pre-defined internal action that is used for inter-agent communication. A list of all pre-defined internal actions is presented in Appendix A.3; because this one is special in that it is the basis for agent interaction in *Jason* – and interaction in multi-agent systems is an essential issue – we need a detailed account of the internal action and all possible uses of it so that programmers can make effective use of agent communication in *Jason*; this is what we do in this chapter. The general form of the pre-defined internal action for communication is:

```
.send(receiver,illocutionary_force,propositional_content)
```

where each parameter is as follows. The `receiver` is simply referred to using the name given to agents in the multi-agent system configuration file. When there are multiple instances of an agent, *Jason* creates individual names by adding numbers starting from 1. For example, if there are three instances of agent a, *Jason* creates agents a1, a2 and a3. The `receiver` can also be a list of agent names; this can be used to multicast the message to the group of agents in the list. The `propositional_content` is a term often representing a literal but at times representing a triggering event, a plan or a list of either literals or plans. The `illocutionary_forces` available are briefly explained below; the name of the performatives and the informal semantics are similar to KQML [61, 69]. The next section will show in detail the changes in an agent's mental state that occur when messages are received. The performatives with suffix 'How' are similar to their counterparts but related to plans rather than predicates.

The list of all available performatives are as follows, where *s* denotes the sender and *r* the receiver:

tell: *s* intends *r* to believe (that *s* believes) the literal in the message's content to be true;

untell: *s* intends *r* not to believe (that *s* believes) the literal in the message's content to be true;

achieve: *s* requests *r* to try and achieve a state of affairs where the literal in the message content is true (i.e. *s* is delegating a goal to *r*);

unachieve: *s* requests *r* to drop the goal of achieving a state of affairs where the message content is true;

askOne: *s* wants to know if the content of the message is true for *r* (i.e. if there is an answer that makes the content a logical consequence of *r*'s belief base, by appropriate substitution of variables);

askAll: *s* wants all of *r*'s answers to a question;

tellHow: *s* informs *r* of a plan (*s*'s know-how);

untellHow: *s* requests that *r* disregard a certain plan (i.e. delete that plan from its plan library);

askHow: *s* wants all of *r*'s plans that are relevant for the triggering event in the message content.

Agents can ignore messages, for example if a message delegating a goal is received from an agent that does not have the power to do so, or from an agent towards which the receiver has no reason to be generous. Whenever a message is received by an agent, if the agent accepts the message, an event is generated which is handled by pre-defined AgentSpeak plans (shown in the next section).

There is another internal action for communication used to broadcast messages rather than send a message to a particular (or a group) of agents. The usage is `.broadcast(illocutionary_force,propositional_content)`, where the parameters are as above, but the message is sent to all agents registered in the agent society (i.e. the *Jason* multi-agent system; note that agents can be created and destroyed dynamically).

6.2 Informal Semantics of Receiving Messages

The effects on a *Jason* agent of receiving a message with each of these types of illocutionary acts is explained (formally) in Section 10.3, following the work in [70, 100]. We here explain this informally, considering the following example. Assume agent s (the 'sender') has just executed the following formula that appeared in one of its intentions:

```
.send(r,tell,open(left_door));
```

agent r would then receive the message

$$\langle s, tell, open(left\_door)\rangle.$$

The first thing that happens when this message is taken from r's mailbox is to pass it as argument in a call to the SocAcc method (it stands for 'socially acceptable'; see Section 7.2). This method can be overridden for each agent to specify which other agents are authorised to send which types of messages. If SocAcc returns false, the message is simply discarded. If the message is accepted, it will be processed by agent r as explained below; the list gives all possible combinations of illocutionary forces and content types. Even though we are explaining what happens when the message is received by r, it makes the presentation easier to use the internal action used by s to *send* the message rather than the message in the format received by r.

We shall now consider each type of message that could be sent by s. We organise the examples in the following categories: information exchange, goal delegation, information seeking and know-how related communication.

Information Exchange

.send(r, tell, open(left_door)): the belief open(left_door) [source(s)] will be added to r's belief base. Recall that this generally means that agent r believes the left door is open because agent s said so, which normally means that s itself believed the left door was open at the time the message was sent and wanted r to believe so too. When programming agents that need to be more conscious of other agents' trustworthiness, the plans need to be programmed so that the semantics of open(left_door)[source(s)] is simply that, at the time the message was received, agent s wanted agent r to believe open(left_door), rather than either of the agents actually believing it.

.send([r1,r2], tell, open(left_door)): the belief open(left_door)[source(s)] will be added to the belief base of both r1 and r2.

.send(r, tell, [open(left_door),open(right_door)]): the belief open(left_door)[source(s)] and the belief open(right_ door)[source(s)] will both be added to r's belief base.

These variations using a list of recipients of having a list in the content field can also be used with all other types of messages. However, note that for a synchronous ask message, the first answer from whatever agent makes the suspended intention be resumed.

.send(r,untell,open(left_door)): the belief open(left_ door)[source(s)] will be removed from r's belief base. Note that if agent r currently believes open(left_door)[source(t), source(s), source(percept)], meaning the agent perceived, and was also told by agents s and t, that the left door is open, the belief base will be updated to have open(left_door)[source(t), source(percept)] instead. That is, only the fact that s wanted r to believe open(left_door) is removed from the belief base.

Figure 6.1 shows a summary of the changes that take place within agents due to communication of type 'information exchange'.

Cycle	s actions	r belief base	r events
1	.send(r, *tell*, open(left_door))		
2		open(left_door) [source(s)]	+open(left_door) [source(s)]
3	.send(r, *untell*, open(left_door))	open(left_door) [source(s)]	
4			-open(left_door) [source(s)]

Figure 6.1 Changes in the receiver's state for (un)tell messages.

Goal Delegation

.send(r, achieve, open(left_door)): the event $\langle +!open(left\_door)[source(s)], \top \rangle$ is created and added to r's set of events. This means that, if the agent has plans that are relevant for when it needs to achieve the state of affairs where the door is open, and one such plan happens to be applicable when the event is selected, that plan will become intended. This in turn means that the agent commits to see to it that the door gets opened. Note how the different illocutionary forces make a significant difference on how the message is interpreted by the agent, as compared with the cases above. Interestingly, *Jason* adds an annotation that the goal was delegated by agent s, so this can be used in the plans to decided whether to actually take any action or not, and to decide what is the most appropriate course of action as well (note that this is on top of the message being socially acceptable).

.send(r, unachieve, open(left_door)): the internal action .drop_desire(open(left_door)) is executed. This internal action is described in Section A.3; it causes current and potential intentions on instances of a goal open(left_door) to be dropped, without generating a plan failure event for the plans that required that goal to be achieved. The default communication plan for this performative could be overridden, for example to use .fail_goal, but then care must be taken to avoid the goal being attempted again within plan failure handling (in case the agent is to follow the request of dropping the goal).

Cycle	s actions	r intentions	r events
1	`.send(r,achieve,` `    open(left_door))`		
2			`+!open(left_door)` `    [source(s)]`
3		`!open(left_door)` `    [source(s)]`	
4	`.send(r,unachieve,` `    open(left_door))`	`!open(left_door)` `    [source(s)]`	
5			

Figure 6.2 Changes in the receiver's state for (un)achieve messages.

Advance

Note that, in the standard interpretation, *all* instances of that goal currently intended by the agent will be dropped. In some applications, it might be useful to only drop the instances of the goal that were delegated by the sender itself. This can be done easily using the source annotations created by *Jason* and by creating a special plan to interpret unachieve messages; we explain later how to create specialised plans for interpreting received messages.

A common situation will be for an agent to ask another agent to drop a goal that itself delegated to that agent in the past. In that case, the goal-addition triggering event appears in the plan just above the pre-defined plan for handling received messages. Although in this case usually nothing needs to be done before the goal is dropped, if necessary in a particular application, there is a chance for a plan to be executed if anything needs to be done before the goal is dropped, by creating a contingency plan for the pre-defined plan used to interpret unachieve messages.

Figure 6.2 shows a summary of the changes that take place within agents due to communication of type 'goal delegation'.

Information Seeking

.send(r, askOne, open(Door), Reply): s's intention that executed this internal action will be suspended until a reply from r is received. On r's side, the message ⟨s, askOne, open(Door)⟩ is received and (if accepted) the *Jason* predefined plan to interpret askOne messages itself sends the appropriate reply; for example, a reply with content `open(left_door)` if that was in r's belief base at the time. This is done with a test goal, so if the relevant information is not in the belief base the agent can also try to execute a plan to obtain the information, if such plan exists in the plan library. When the reply is received by s, the content in that message is instantiated to the `Reply` variable in the internal action executed by s. The intention is then resumed, hence the formula after the `.send` internal action is only executed after s has received a reply from r. Interestingly, this means that an `askOne` message is similar to a test goal but in another agent's belief base (if the other agent so permits). Further, as mentioned earlier, recall that if a synchronous `ask` message is multi-cast, the first reply received from any of the addressees of the original message is used to resume the suspended intention.

Note that variable `Door` in this internal action would *not* be instantiated after the action is executed, only the *Reply* variable is. To obtain a similar effect of instantiating the `Door` variable as if the reply was obtained from the message content itself (the third parameter), the action needs to be written as '`.send(r,askOne,open(Door), open(Door))`'. This looks redundant but is necessary for technical reasons, e.g. to allow for negative replies and for communication with agents that are not written in AgentSpeak: the content and reply can be any AgentSpeak term, including a string, which is then general enough to allow communication among heterogeneous agents.

If the content of s's ask message does not unify with any of r's beliefs, variable `Reply` would be unified with the reserved keyword `false`. So, to write a plan that can only carry on executing if another agent does *not* believe `open(Door)` (i.e. that it does not believe that any door is open), we include '`.send(r,ask,open(Door),false);`' in the plan body. In this case, if r believes the left door is open, the reply would be `open(left_door)`, which cannot be unified with `false`, therefore the internal action `.send` fails, and so does the plan where the action appeared. Note that checking whether the receiver believes an instance of the propositional content uses a *test goal*, therefore before a `false` answer is sent, an event `+?open(Door)` would be generated, and if the agent had a relevant (and applicable) plan, that plan would be executed and could then lead to the answer to be sent to the agent waiting for a reply to an `ask` message.

One frequent mistake is to think that, because '`ask`' messages are synchronous, agents become idle waiting for any reply. Recall that an AgentSpeak agent typically has multiple foci of attention (i.e. multiple intentions in the set of intentions) and so the agent can carry on doing work for the intentions that do not need to wait for an answer. Also, asynchronous requests for information can be done by providing the receiver with a plan to achieve the goal of providing the required information, and the sender can have a plan which simply sends an `achieve` message with that particular goal as message content.

Advance

While there is no problem with the agent becoming completely idle, there is, however, a problem in that agents could end up with many suspended intentions which, for example, could be dropped as they will never be successfully completed (e.g. because the relevant agent left the society, or failed to receive the message). To help alleviate the problem, any 'ask' message can have an optional fifth parameter which determines a 'timeout' for the wait for a reply (in milliseconds), for example as in '`.send(r,askOne,open(Door),Reply,3000)`' to wait for a reply within 3 seconds. If the request for information times out, the `Reply` variable will be instantiated with the term `timeout`, so that programmers will know that no reply was obtained; this can be used by programmers in a contingency plan to, for example, 'give up' by dropping the intention, or 'insist' by sending the message again.

`.send(r, askOne, open(Door))`: as above but asynchronous, in the sense that the `askOne` message is sent to `r` but the intention is not suspended to wait for a reply; this means that the formula in the plan body immediately after the `.send` internal action could in principle execute in the next reasoning cycle (if selected for execution). Of course this should be used when the remainder of the plan does not depend on the reply being received to finish executing. When a reply is received, a new belief is likely to be added which, recall, generates an event. When further action is necessary if a reply eventually arrives, programmers then typically write plans to deal, in a separate intention, with the event generated by the received response.

`.send(r, askAll, open(Door), Reply)`: as for `askOne` (the synchronous version) but `Reply` is a list with all possible answers that `r` has for the content. For example, if at the time of processing the `askAll` message agent `r` believed both `open(left_door)` and `open(right_door)`, variable `Reply` in the internal action would be instantiated with the list '`[open(left_door),open(right_door)]`'.

An empty list as reply denotes r did not believe anything that unifies with open(Door).

Figure 6.3 shows a summary of the changes that take place within agents due to communication of type 'information seeking'. Note that this figure is for the

		r belief base
		open(left_door)
		open(right_door)

Cycle	s actions / unifier / events	r actions
1	.send(r,askOne, open(Door),Reply)	
2		.send(s,tell,open(left_door))
3	Reply ↦ open(left_door)	
4	.send(r,askOne, closed(Door),Reply)	
5		.send(s,tell,false)
6	Reply ↦ false	
7	.send(r,askOne, open(Door))	
8		.send(s,tell,open(left_door))
9	+open(left_door)[source(s)]	
10	.send(r,askAll, open(D),Reply)	

11		.send(s,tell,
		    [open(left_door),
		    open(right_door)])

12	Reply ↦ [open(left_door), open(right_door)]	
13	.send(r,askAll, open(D))	

14		.send(s,tell,
		    [open(left_door),
		    open(right_door)])

15	+open(left_door)[source(s)] +open(right_door)[source(s)]	

Figure 6.3 Examples of ask protocols.

default `ask` protocol assuming no relevant plans with triggering events +? are available in the receiver agent.

Know-how Related

`.send(r, tellHow, "@pOD +!open(Door) : not locked(Door) <- turn_handle(Door); push(Door); ?open(Door).")`: the string in the message content will be parsed into an AgentSpeak plan and added to `r`'s plan library. Note that a string representation for a plan in the agent's plan library can be obtained with the `.plan_label` internal action given the label that identifies the required plan; see page 243 for details on the `.plan_label` internal action.

`.send(r, tellHow, ["+!open(Door) : locked(Door) <- unlock(Door); !!open(Door).", "+!open(Door) : not locked(Door) <- turn_handle(Door); push(Door); ?open(Door)."])`: each member of the list must be a string that can be parsed into a plan; all those plans are added to `r`'s plan library.

`.send(r, untellHow, "@pOD")`: the string in the message content will be parsed into an AgentSpeak plan label; if a plan with that label is currently in `r`'s plan library, it is removed from there. Make sure a plan label is explicitly added in a plan sent with `tellHow` if there is a chance the agent might later need to tell other agents to withdraw that plan from their belief base.

`.send(r, askHow, "+!open(Door)")`: the string in the message content must be such that it can be parsed into a triggering event. Agent `r` will reply with a list of all *relevant plans* it currently has for that particular triggering event. In this example, `s` wants `r` to pass on his know-how on how to achieve a state of affairs where something is opened (presumably doors, in this example). Note that there is not a fourth argument with a 'reply' in this case, differently from the other 'ask' messages. The plans sent by `r` (if any) will be automatically added to `s`'s plan library. In Section 11.2, we explain a more elaborate approach to plan exchange among collaborating agents.

The AgentSpeak Plans for Handling Communication

Advance

As we mentioned earlier, instead of changing the interpreter to allow for the interpretation of received speech-act based messages, we have written AgentSpeak plans to interpret received messages; *Jason* includes such plans at the end of the plan library when any agent starts running. There are separate plans to handle each type of illocutionary force, and in some cases to handle different situations (e.g. whether the agent has an answer for an 'ask' message or not). The advantage of the approach of using plans to interpret messages is that the normal event/intention selection functions can be used to give appropriate priority to communication requests, and also this allows users to customise the way messages are interpreted easily, by just including in their AgentSpeak code new versions of the pre-defined communication plans. Because the user-defined plans will appear first in the plan library, they effectively override the pre-defined ones. The predefined plans are shown below, and they can also be found in file `src/asl/kqmlPlans.asl` of *Jason*'s distribution.

By now, readers will be able to understand the plans, as they are rather simple, so we do not provide detailed descriptions. One interesting thing to note, though, is the heavy use of higher-order variables – variables instantiated with a belief, goal, etc. rather than a term (see the use of formulæ such as +CA and !CA in the plans below). Another noticeable feature of the plans is the use of various internal actions: as some of these have not yet been introduced, readers aiming to understand these plans in detail might need to refer back an forth to Appendix A.3, where we describe in detail all the predefined internal actions in *Jason* (even though the action names will give a good intuition of what they do). Also, note that the names of variables used are long and unusual to ensure they do not conflict with user variables in the message content.

```
// Default plans to handle KQML performatives
// Users can override them in their own AS code
//
// Variables:
//    KQML_Sender_Var:  the sender (an atom)
//    KQML_Content_Var: content (typically a literal)
//    KQML_MsgId:       message id (an atom)
//


/* ---- tell performatives ---- */

@kqmlReceivedTellStructure
+!kqml_received(KQML_Sender_Var, tell, KQML_Content_Var, KQML_MsgId)
   :  .structure(KQML_Content_Var) &
      .ground(KQML_Content_Var) &
      not .list(KQML_Content_Var)
   <- .add_annot(KQML_Content_Var, source(KQML_Sender_Var), CA);
      +CA.
@kqmlReceivedTellList
+!kqml_received(KQML_Sender_Var, tell, KQML_Content_Var, KQML_MsgId)
   :  .list(KQML_Content_Var)
   <- !add_all_kqml_received(KQML_Sender_Var,KQML_Content_Var).
```

```
@kqmlReceivedTellList1
+!add_all_kqml_received(_,[]).

@kqmlReceivedTellList2
+!add_all_kqml_received(S,[H|T])
    :  .structure(H) &
       .ground(H)
   <- .add_annot(H, source(S), CA);
      +CA;
      !add_all_kqml_received(S,T).

@kqmlReceivedTellList3
+!add_all_kqml_received(S,[_|T])
   <- !add_all_kqml_received(S,T).

@kqmlReceivedUnTell
+!kqml_received(KQML_Sender_Var, untell, KQML_Content_Var, KQML_MsgId)
   <- .add_annot(KQML_Content_Var, source(KQML_Sender_Var), CA);
      -CA.


/* ---- achieve performatives ---- */

@kqmlReceivedAchieve
+!kqml_received(KQML_Sender_Var, achieve, KQML_Content_Var, KQML_MsgId)
   <- .add_annot(KQML_Content_Var, source(KQML_Sender_Var), CA);
      !!CA.

@kqmlReceivedUnAchieve[atomic]
+!kqml_received(KQML_Sender_Var, unachieve, KQML_Content_Var, KQML_MsgId)
   <- .drop_desire(KQML_Content_Var).


/* ---- ask performatives ---- */

@kqmlReceivedAskOne1
+!kqml_received(KQML_Sender_Var, askOne, KQML_Content_Var, KQML_MsgId)
   <- ?KQML_Content_Var;
      .send(KQML_Sender_Var, tell, KQML_Content_Var, KQML_MsgId).

@kqmlReceivedAskOne2 // error in askOne, send untell
-!kqml_received(KQML_Sender_Var, askOne, KQML_Content_Var, KQML_MsgId)
   <- .send(KQML_Sender_Var, untell, KQML_Content_Var, KQML_MsgId).

@kqmlReceivedAskAll1
+!kqml_received(KQML_Sender_Var, askAll, KQML_Content_Var, KQML_MsgId)
    :  not KQML_Content_Var
   <- .send(KQML_Sender_Var, untell, KQML_Content_Var, KQML_MsgId).

@kqmlReceivedAskAll2
+!kqml_received(KQML_Sender_Var, askAll, KQML_Content_Var, KQML_MsgId)
   <- .findall(KQML_Content_Var, KQML_Content_Var, List);
      .send(KQML_Sender_Var, tell, List, KQML_MsgId).
```

```
/* ---- know-how performatives ---- */

// In tellHow, content must be a string representation
// of the plan (or a list of such strings)

@kqmlReceivedTellHow
+!kqml_received(KQML_Sender_Var, tellHow, KQML_Content_Var, KQML_MsgId)
   <- .add_plan(KQML_Content_Var, KQML_Sender_Var).

// In untellHow, content must be a plan's
// label (or a list of labels)
@kqmlReceivedUnTellHow
+!kqml_received(KQML_Sender_Var, untellHow, KQML_Content_Var, KQML_MsgId)
   <- .remove_plan(KQML_Content_Var, KQML_Sender_Var).

// In askHow, content must be a string representing
// the triggering event
@kqmlReceivedAskHow
+!kqml_received(KQML_Sender_Var, askHow, KQML_Content_Var, KQML_MsgId)
   <- .relevant_plans(KQML_Content_Var, ListAsString);
      .send(KQML_Sender_Var, tellHow, ListAsString, KQML_MsgId).
```

The plans above are loaded by *Jason* at the end of any AgentSpeak program; users can customise the interpretation of received speech-act based messages by creating similar plans (which can appear anywhere in the program[2]). Whilst it can be useful to customise these plans for very specific applications, much care should be taken in doing so, to avoid changing the semantics of communication itself. If a programmer chooses not to conform to the semantics of communication as determined by the pre-defined plans (and the literature referred to earlier), sufficient documentation must be provided. Before doing so, it is important to ensure it is necessary not to conform to the given semantics; there are often ways of getting around specific requirements for communication without changing the semantics of communication. An example of a situation where simple customisation can be useful is as follows. Suppose an agent needs to remember that another agent asked it to drop a particular goal. Plan @kqmlReceivedUnAchieve could be changed so that the plan body includes +askedToDrop(KQMLcontentVar)[source(KQML_Sender_Var)], possibly even with another annotation stating when that occurred. In other cases, it might be useful to change that plan so that an agent *a* can only ask another agent *b* to drop a goal

---

[2]This assumes the applicable plan selection function has not been overridden, so applicable plans are selected in the order they appear in the program.

that was delegated by *a* itself. This could be easily done by adding this plan to the source code for agent *b*:

```
@kqmlReceivedUnAchieve[atomic]
+!kqml_received(KQMLsenderVar, unachieve, KQMLcontentVar,
    KQMLmsgId) <- .add_annot(KQMLcontentVar, source(
        KQMLsenderVar), KQMLannotG); .drop_desire(KQMLannotG).
```

Recall that the default handling of `achieve` always adds an annotation specifying which was the agent that delegated the goal.

6.3   Example: Contract Net Protocol

This example shows how communication support available in *Jason* helps the development of the well known Contract Net Protocol (CNP) [36, 90]. We show an implementation in AgentSpeak of the version of the protocol, as specified by FIPA [45]. In this protocol one agent, called 'initiator', wishes to have some task performed and asks other agents, called 'participants', to bid to perform that task. This asking message is identified by `cfp` (call for proposals). Some of the participants send their proposals or refusals to the initiator. After the deadline has passed, the initiator evaluates the received proposals and selects one agent to perform the task. Figure 6.4 specifies the sequence of messages exchanged between the agents in a CNP interaction.

The project for this example determines that it is composed of six agents: one initiator and five participants.

```
MAS cnp {
    agents:
        c;      // the CNP initiator
        p #3;   // three participants able to send proposals
        pr;     // a participant that always refuses
        pn;     // a participant that does not answer
}
```

Participant Code

Three types of agents play the participant role in this example of the use of the CNP. Once started, all participants send a message to the initiator (identified by the name c) introducing themselves as 'participants'.

The simpler agent of the application is named `pn` and this agent has only one plan used to introduce itself as a participant. When its initial belief

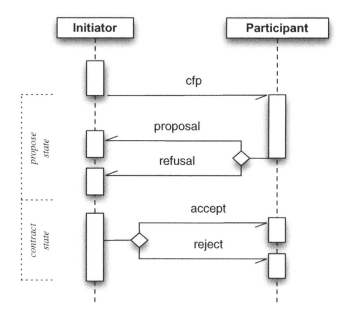

Figure 6.4 Contract net protocol.

`plays(initiator,c)` is added in the belief base, the event `+plays(initiator,c)` is added in the event queue and the respective plan executed. Although it says to the initiator that it is a participant, this agent never sends proposals or refusals.

```
// the name of the agent playing initiator in the CNP
plays(initiator,c).

// send a message to the initiator introducing myself
// as a participant
+plays(initiator,In)
   :   .my_name(Me)
   <- .send(In,tell,introduction(participant,Me)).
```

The second type of participant rejects all calls for proposals. When this agent receives a message from c informing it that a CNP is starting, it ignores the content and just sends a refusal message. Each `cfp` message has two arguments: a unique identification of the conversation (an 'instance' of the protocol) so that agents can participate in several simultaneous CNPs, and a description of the task being contracted. In this application, the `cfp` message is sent by the initiator using the `tell` performative; a belief such as `cfp(1,fix(computer))[source(c)]` is therefore added to the receiver's belief base and the corresponding event `+cfp(1,fix(computer))[source(c)]` added to their event queue. The agent's plan library has a plan, shown below, to react to this event.

```
plays(initiator,c).
+plays(initiator,In)
    :  .my_name(Me)
    <- .send(In,tell,introduction(participant,Me)).


// plan to answer a CFP
+cfp(CNPId,Task)[source(A)]
    :   plays(initiator,A)
    <- .send(A,tell,refuse(CNPId)).
```

Based on the project definition, *Jason* creates three instances of the third type of participant (named p). The proposals sent by these agents is formed with their price for the task execution. However, in this example, the price is randomly defined by the first rule in the code. This rule is used to answer the cfp message in plan @c1. The agent also has two plans (@r1 and @r2) to react to the result of the CNP.

```
// gets the price for the product,
// (a random value between 100 and 110).
price(Task,X) :- .random(R) & X = (10*R)+100.


plays(initiator,c).


/* Plans */


// send a message to initiator introducing myself
// as a participant
+plays(initiator,In)
    :  .my_name(Me)
    <- .send(In,tell,introduction(participant,Me)).


// answer a Call For Proposal
@c1 +cfp(CNPId,Task)[source(A)]
    :  plays(initiator,A) & price(Task,Offer)
    <- +proposal(CNPId,Task,Offer); // remember my proposal
       .send(A,tell,propose(CNPId,Offer)).


@r1 +accept_proposal(CNPId)
    :  proposal(CNPId,Task,Offer)
    <- .print("My proposal '",Offer,"' won CNP ",CNPId,
             " for ",Task,"!").
       // do the task and report to initiator
```

```
@r2 +reject_proposal(CNPId)
   <- .print("I lost CNP ",CNPId, ".");
      -proposal(CNPId,_,_). // clear memory
```

Initiator Code

The behaviour of the initiator agent can be briefly described by the following steps: start the CNP, send a `cfp`, wait for the answers, choose the winner, and announce the result. The complete code of this agent is as follows and is explained below.

```
/* Rules */

all_proposals_received(CNPId) :-
  .count(introduction(participant,_),NP) & // number of participants
  .count(propose(CNPId,_), NO) &      // number of proposes received
  .count(refuse(CNPId), NR) &         // number of refusals received
  NP = NO + NR.

/* Initial goals */

!startCNP(1,fix(computer_123)).

/* Plans */

// start the CNP
+!startCNP(Id,Object)
   <- .wait(2000);  // wait participants introduction
      +cnp_state(Id,propose);   // remember the state of the CNP
      .findall(Name,introduction(participant,Name),LP);
      .print("Sending CFP to ",LP);
      .send(LP,tell,cfp(Id,Object));
      .concat("+!contract(",Id,")",Event);
      // the deadline of the CNP is now + 4 seconds, so
      // the event +!contract(Id) is generated at that time
      .at("now +4 seconds", Event).


// receive proposal
// if all proposal have been received, don't wait for the deadline
@r1 +propose(CNPId,Offer)
   :  cnp_state(CNPId,propose) & all_proposals_received(CNPId)
   <- !contract(CNPId).

// receive refusals
@r2 +refuse(CNPId)
```

```
    :   cnp_state(CNPId,propose) & all_proposals_received(CNPId)
    <- !contract(CNPId).

// this plan needs to be atomic so as not to accept
// proposals or refusals while contracting
@lc1[atomic]
+!contract(CNPId)
    :   cnp_state(CNPId,propose)
    <- -+cnp_state(CNPId,contract);
        .findall(offer(O,A),propose(CNPId,O)[source(A)],L);
        .print("Offers are ",L);
        L \== []; // constraint the plan execution to at least one offer
        .min(L,offer(WOf,WAg)); // sort offers, the first is the best
        .print("Winner is ",WAg," with ",WOf);
        !announce_result(CNPId,L,WAg);
        -+cnp_state(Id,finished).

// nothing todo, the current phase is not 'propose'
@lc2 +!contract(CNPId).

-!contract(CNPId)
    <- .print("CNP ",CNPId," has failed!").

+!announce_result(_,[],_).
// announce to the winner
+!announce_result(CNPId,[offer(O,WAg)|T],WAg)
    <- .send(WAg,tell,accept_proposal(CNPId));
        !announce_result(CNPId,T,WAg).
// announce to others
+!announce_result(CNPId,[offer(O,LAg)|T],WAg)
    <- .send(LAg,tell,reject_proposal(CNPId));
        !announce_result(CNPId,T,WAg).
```

The agent has the initial goal !startCNP(1,fix(...)), where 1 means the CNP instance identification and fix(...) the task being constructed. The intention created by the !startCNP goal is initially suspended for 2 seconds while the participants send their own introductions. When resumed, the intention adds a belief to remember the current state of the protocol: the initial state is propose. This belief constrain the receiving of proposal and refusals (see plans @r1 and @r2). Based on the introduction beliefs received in the introduction phase, the plan builds a list of all participants' names and send the cfp to them. Note that we can inform a *list* of receivers in the .send internal action (i.e. .send allows for multicast).

The initiator starts to receive proposals and refusals and plans @r1 and @r2 check whether all participants have answered and the protocol is still in the `propose` state. If the initiator has all answers, it can go on and contract the best offer by means of the goal `!contract`. However it could be the case that not all participants answered the `cfp`, and in such case the belief `all_proposals_received(CNPId)` never holds. To deal with this problem, before finishing the intention for `!startCNP`, the agent uses the internal action `.at` to generate the event `+!contract(1)` 4 seconds after the protocol has started. If the initiator receives all answers before 4 seconds, this event is handled by the plan `@lc2`, and so discarded. Note that plan `@lc1` is `atomic`: when it starts executing, no other intention is selected for execution before it finishes. The first action of the plan is to change the protocol state, which is also used in the context of the plan. It ensures that the intention for the goal `!contract` is never performed twice. Without this control, the goal `!contract` could be triggered more than once for the same CNP instance, by plans `@r1/@r2` and also by the event generated by the `.at` internal action.

A run of this *Jason* application could result in an output as follows:

```
[c]   saying: Sending CFP to [p2,p3,pr,pn,p1]
[c]   saying: Offers are [offer(105.42,p3),
                          offer(104.43,p2),
                          offer(109.80,p1)]
[c]   saying: Winner is p2 with 104.43
[p3]  saying: I lost CNP 1.
[p1]  saying: I lost CNP 1.
[p2]  saying: My proposal '104.43' won CNP 1
              for fix(computer\_123)!
```

6.4 Exercises

1. Considering the domestic robot scenario from Chapter 5:

Basic
   (a) Change the AgentSpeak code for the robot so that a message is sent to the supermarket to inform it that a delivery of beer arrived.

Basic
   (b) Create a new supermarket agent (called nsah) that does not have a plan for goal `+!order(P,Q)`, but once started, uses the performative `askHow` to ask another supermarket for such a plan.

   Rewrite the program for the supermarket agent that sent the plan so that after 10 seconds from the start of the application it sends an `untellHow` message to nsah.

vanced
   (c) Instead of each supermarket sending the beer price to the robot as soon as it starts (as proposed in the last exercise of Section 3.5), change the code for those agents so that is uses the contract net protocol in such

way that the robot plays the initiator role and the supermarkets play the participant role.

(d) Create a new supermarket agent which, when asked by the robot for a proposal, pretends to be a domestic robot itself and asks proposals from other supermarkets also using the CNP (i.e. it plays the initiator role). With the best offer from the others, this supermarket subtracts 1% from the price and, provided that price still gives the supermarket some profit, sends this new value as its own proposal, thereby probably winning the actual CNP initiated by the domestic robot. `Advance`

(e) Create two instances of the previous supermarket agent. This causes a loop in the CNP! Propose a solution for this problem. `Advance`

2. Suppose the initiator of a CNP may want to cancel the cfp. In order to implement this feature, add a plan in the initiator program for events such as +!abort(CNPId) that uses the untell performative to inform the participants that the CNP was cancelled. On the participants' side, write plans to handle this untell message (this plan basically removes the offer from the belief base). `Basic`

3. Assume now that participants may also cancel their offer. Write plans (on both sides) to deal with this situation. Note that a participant can cancel its offer only if the protocol state is propose. Otherwise the initiator should inform the participant that its proposal was *not* cancelled. `Basic`

4. Another approach to implement the CNP is to use the askOne performative. In this case, the initiator code is as follows (only the plans that need to be changed are shown below). `Advance`

```
+!startCNP(Id,Task)
   <- .wait(2000);  // wait for participants' introduction
      .findall(Name,introduction(participant,Name),LP);
      .print("Sending CFP to ",LP);
      !ask_proposal(Id,Object,LP).

/** ask proposals from all agents in the list */

// all participants have been asked, proceed with contract
+!ask_proposal(CNPId,_,[])
   <- !contract(CNPId).

// there is a participant to ask
+!ask_proposal(CNPId,Object,[Ag|T])
   <- .send(Ag,
           askOne,
           cfp(Id,Object,Offer),
```

```
                      Answer,
                      2000); // timeout = 2 seconds
               !add_proposal(Ag,Answer); // remember this proposal
               !ask_proposal(CNPId,Object,T).

        +!add_proposal(Ag,cfp(Id,_,Offer))
           <- +offer(Id,Offer,Ag).  // remember Ag's offer
        +!add_proposal(_,timeout).   // timeout, do nothing
        +!add_proposal(_,false).     // answer is false, do nothing

        +!contract(CNPId) : true
           <- .findall(offer(O,A),offer(CNPId,O,A),L);
              .print("Offers are ",L);
              L \== [];
              .min(L,offer(WOf,WAg));
              .print("Winner is ",WAg," with ",WOf);
              !announce_result(CNPId,L,WAg).
```

The code for the participants also needs changing because no belief is added
when receiving an ask message; instead, the required `price` information is
directly retrieved from the belief base with a plan for a test goal.

```
// in place of +cfp(....)
+?cfp(CNPId,Object,Offer)
    :  price(Object,Offer)
    <- +proposal(CNPId,Object,Offer). // remember my proposal
```

Compare the two implementations of the CNP and identify advantages and
disadvantages of the different approaches.

vanced 5. Write AgentSpeak plans that agents can use to reach *shared beliefs* under
the assumption that the underlying communication infrastructure guarantees the delivery of all messages and that neither agent will crash before
messages are received. In *Jason*, we can say two agents `i` and `j` share a
belief `b` whenever `i` believes `b[source(self),source(j)]` and `j` believes
`b[source(self),source(i)]`. Typically, one of the agents will have the
initiative (i.e. goal) of reaching a shared belief with another agent. Ideally,
you should write plans so that the same plans can be used in both the agent
taking the initiative as well as its interlocutor who will have to respond to
the request to achieve a shared belief.

vanced 6. When an agent sends an achieve message to another agent, it does not get
any feedback. It is thus unable to know whether or when the task has been
accomplished. Create a new performative `achievefb` so that the sender will
receive a feedback message, as in the following example:

```
!somegoal
    <- .send(bob,achievefb,go(10,20)).
```

```
+achieved(go(X,Y))[source(bob)]
    <- .print("Hooray, Bob finished the job!").
+unachieved(go(X,Y))[source(bob)]
    <- .print("Damn it, Bob didn't do as told.").
```

You could also write plans for this agent to 'put pressure' on Bob, if it is taking too long for Bob to achieve the goal (or the agent realises again the need to have achieved the goal delegated to Bob). Of course, Bob would also need to react appropriately to the pressure (having plans for that), otherwise those messages would just be ignored.

Note that, while the default communication plans are in the kqmlPlans.asl file, you can always add plans for communication, similar to the default communication plans, to your agent code (your plans for a particular performative will override the default plans for that performative). Needless to say, if you plan to change the semantics of communication, you should be extra careful when doing this (and ask yourself if it is worth doing); however, in some cases this might be necessary (e.g. if you want your agent to be able to lie). It also worth mentioning that, strictly speaking, we do not need a new performative for the 'feedback' exercise; we just need to make sure the agents implement suitable protocols, so an achieve performative leads to confirmation whenever necessary in the given application. On the other hand, the advantage of having two different performatives is that they can be used to make it clear whether the requesting agent expects confirmation or not for a particular request.

7

User-Defined Components

Jason is distributed with a basic set of functionalities, as described in the previous chapters. However, developers will typically have a variety of specific requirements when developing their systems, such as, for example, database access, legacy system integration, graphical interfaces, custom agent perception/capabilities, special intention handling, and many others. To fulfil these requirements, an extensibility-based approach has been adopted in the design of *Jason*. As *Jason* itself is developed in Java, such customisations and extensions will involve some Java programming. It is possible to develop agents with a fully declarative language as AgentSpeak and also to extend the language with libraries and components developed using the more traditional object-oriented paradigm. New commands can be added to the language by means of user-defined internal actions and several components of the interpreter can be customised. New internal actions as well as customised components are programmed in Java, so some knowledge of that programming language is required to use these features. Referring to the *Jason* API documentation, available in the distribution, is also very useful when programming new user-defined components.

There are two main reasons for creating an internal action: to extend the agent internal capabilities when object orientation is a better abstraction level for the required programming task than that provided by AgentSpeak; and to allow the agent to use legacy code already programmed in Java or other languages (using Java Native Interface). There are also *standard internal actions* which are those available as part of the *Jason* distribution. Some standard internal actions have already appeared throughout the text, such as .print, .send and .my_name (a complete list is presented in Appendix A.3). The next section concentrates on how *new* internal actions can be created and used in an agent code.

Programming Multi-Agent Systems in AgentSpeak using Jason   R.H. Bordini, J.F. Hübner, M. Wooldridge
© 2007 John Wiley & Sons, Ltd

Besides extending the set of internal actions to be used by the agents, developers can also customise and extend different components of the interpreter. Section 7.2 describes how the *Agent* class can be customised; this class determines part of the functioning of the interpreter such as the belief-revision and selection functions used in the reasoning cycle. Section 7.3 describes the *Agent Architecture* customisation; this architecture defines how the agent interacts with the external world, i.e. the environment and other agents. In Section 7.4 it is described how to customise an agent's *belief base*. Finally, Section 7.5 describes how the parsing can be extended by means of user-defined *pre-processing directives*.

7.1   Defining New Internal Actions

User-defined internal actions should be organised in specific libraries, as described below. In the AgentSpeak code, an Internal Action (IA) is accessed by the name of the library, followed by '.', followed by the name of the action. For instance, a new internal action called distance within a library called math, can be used both in the context and in the body of AgentSpeak plans as in the following code:

```
+event : true <- math.distance(10,10,20,30,D); ...
+event : math.distance(10,10,20,30,D) & D > 30 <- ...
```

Libraries are defined as Java packages and each action in the user library should be a Java class within that package; the names of the package and class are the names of the library and action as they will be used in AgentSpeak programs. Recall that all identifiers starting with an uppercase letter in AgentSpeak denote variables, so the name of the library and the internal action class *must* start with a lowercase letter.[1]

All classes defining internal actions should implement the InternalAction interface (see Figure 7.1). There is a default implementation of this interface, called DefaultInternalAction, that simplifies the creation of new internal actions. For example, the declaration of a Java class that implements an internal action to calculate the 'Manhattan distance' between two points is:

```
package math;

import jason.*;
import jason.asSyntax.*;
import jason.asSemantics.*;

public class distance extends DefaultInternalAction {
```

---

[1] The *Jason* IDE provides wizards to create new internal actions in a MAS project and to compile them. These issues are therefore not detailed here.

```
@Override
public Object execute( TransitionSystem ts,
                       Unifier un,
                       Term[] args ) throws Exception {

    < the code that implements the IA goes here >

    }
}
```

As expected, `math` in the example above is the name of the library, and `distance` is the name of the particular action being defined in this Java file. Internal actions should override the **execute** method. This method is called by the AgentSpeak interpreter to execute the internal action. The first argument of this method is the transition system (as defined by the operational semantics of AgentSpeak), which contains all information about the current state of the agent being interpreted (see the *Jason* API for more details about this class). The second is the unifying function currently determined by the execution of the plan where the internal action appeared, or the checking of whether the plan is applicable; this depends on whether the internal action being executed appeared in the body or the context of a plan. The unifying function is important in case the value bound to AgentSpeak variables needs to be used in the implementation of the action. The third argument is an array of terms and contains the arguments given to the internal actions by the user in the AgentSpeak code that called the internal action. The second method of the DefaultInternalAction class is called suspendIntention; this method should return true when the IA causes the intention to be suspended state, as `.at` and also `.send` with `askOne` as performative. Its default implementation in class DefaultInternalAction returns false.

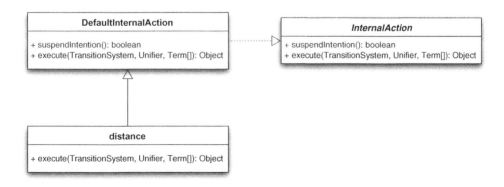

Figure 7.1 Internal action class diagram.

Note that the user needs to make sure the intention will be resumed when appropriate if the **suspendIntention** method is overridden.

The Java code in the **execute** method for the `math.distance` internal action has three steps:

```
try {
    // 1. gets the arguments as typed terms
    NumberTerm p1x = (NumberTerm)args[0];
    NumberTerm p1y = (NumberTerm)args[1];
    NumberTerm p2x = (NumberTerm)args[2];
    NumberTerm p2y = (NumberTerm)args[3];

    // 2. calculates the distance
    double r = Math.abs(p1x.solve()-p2x.solve()) +
               Math.abs(p1y.solve()-p2y.solve());

    // 3. creates the term with the result and
    //    unifies the result with the 5th argument
    NumberTerm result = new NumberTermImpl(r);

    return un.unifies(result,args[4]);

} catch (ArrayIndexOutOfBoundsException e) {
    throw new JasonException("The internal action 'distance'"+
                        "has not received five arguments!");
} catch (ClassCastException e) {
    throw new JasonException("The internal action 'distance'"+
            "has received arguments that are not numbers!");
} catch (Exception e) {
    throw new JasonException("Error in 'distance'");
}
```

The first step is to get references for the four numbers given as parameters for which the distance will be calculated. The type of these objects is **NumberTerm**, which is an interface implemented by **NumberTermImpl** (a single number), **VarTerm** (a variable that may be bound to a number) or **ArithExpr** (an arithmetic expression such as 'X*2' which can be evaluated to a number).

The second step simply computes the distance value. Note that the method `solve()` is used the get the numeric value of the **NumberTerm** object, so that expressions can be used as parameters, as in `math.distance(X1+2,Y1/5-1,20,20,D)`.

In the last step, a **NumberTerm** object representing the result is created. This object is then unified with the fifth argument. This unification may fail in a case such as `math.distance(1,1,2,2,50)`, or succeed in cases such as `math.distance(1,1,2,2,D)` with D free or bound to 2. The result of this unification is also the result of the IA execution. In the case where the IA is used in the context of a plan, returning false means that the plan is not applicable (assuming the context is a simple conjunction where the internal action appeared). When used in a plan body, returning false will fail the plan that called the IA.

Looking at the code of the standard internal actions available in *Jason*'s distribution is often very useful for users trying to create their own internal actions.

vanced

Internal actions and backtracking

An IA may return either a boolean value as in the above example or an iterator of unifiers. In the first case, the boolean value represents whether the execution of the IA was successful or not. The second case is used to backtrack alternative solutions when the IA is used in the context of plans. For example, suppose one needs to write an IA that produces all odd numbers; this IA should return in subsequent calls a sequence of possible unifications such as $\{x \mapsto 1\}, \{x \mapsto 3\}, \ldots$. However, the IA should also work for testing if the argument is an odd number, in which case it should return a boolean answer as usual. Such an IA could then be used in plans such as:

```
+event : math.odd(X) & X > 10 <- ...
+event(X) : math.odd(X) <- ...
```

Note that the plan selection should backtrack for the initial possible values for x in the first plan. In the second plan, assuming that x is ground, the IA only needs to check whether x's value is odd. The code for this IA is:

```
package math;

import jason.asSemantics.*;
import jason.asSyntax.*;
import java.util.*;

public class odd extends DefaultInternalAction {

    @Override
    public Object execute(TransitionSystem ts,
                          final Unifier un,
                          final Term[] arg) throws Exception {

        if (! arg[0].isVar()) {
            // the argument is not a variable, single answer
            if (arg[0].isNumeric()) {
                NumberTerm n = (NumberTerm)arg[0];
                return n.solve() % 2 == 1;
```

```
                    } else {
                       return false;
                    }

               } else {

                   // returns an iterator of unifiers,
                   // each unifier has the arg[0] (a variable)
                   // assigned to an odd number.

                   return new Iterator<Unifier>() {
                       int last = 1;

                       // we always have a next odd number
                       public boolean hasNext() { return true; }

                       public Unifier next() {
                           Unifier c = (Unifier)un.clone();
                           c.unifies(new NumberTermImpl(last), arg[0]);
                           last += 2;
                           return c;
                       }

                       public void remove() {}
                   };
       } } }
```

For another example of such kind of IA, see the source code of the standard IA .member, available in the *Jason* distribution.

Another issue about IAs is that each agent creates an instance of the internal action class the first time that it is executed and reuses the same instance in all subsequent executions of that internal action by that agent; this means that an internal action can be stateful *within the agent*.

7.2   Customising the Agent Class

From the point of view of an (extended) AgentSpeak interpreter as described in Chapter 4, an *agent* is a set of beliefs, a set of plans, some user-defined selection functions and trust function (a 'socially acceptable' relation for received messages), the Belief Update Function, the Belief Revision Function and a 'circumstance' which includes the pending events, intentions and various other structures that are necessary during the interpretation of an AgentSpeak agent. The default

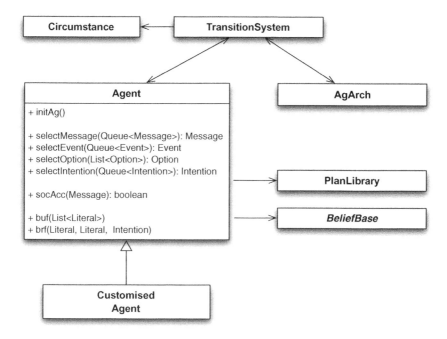

Figure 7.2 Agent class diagram.

implementation of these functions is coded in a class called Agent, which can be customised by developers in order to extend the basic functionalities. The class diagram in Figure 7.2 briefly presents this class and those related to it; however, the diagram does not contain all methods of the class, only the ones that are relevant for the material covered in this chapter. The interested reader is referred to the API documentation for a more detailed description of the classes. The methods of the Agent class that are normally overridden are:

- selectMessage(Queue<Message> mailbox): selects the message that will be handled in the current reasoning cycle; it implements the S_M function in Figure 4.1. The default implementation removes and returns the first message in the agent's mail box.

- selectEvent(Queue<Event> events): selects the event that will be handled in the current reasoning cycle; it implements the $S_{\mathcal{E}}$ function in Figure 4.1. The default implementation removes and returns the first event in the queue.

- selectOption(List<Option> options): this method is used to select one among several options (an applicable plan and an unification) to handle an event. It implements the S_O function in Figure 4.1. The default implementation removes and returns the first option according to the order in which plans were written in the agent code, and the first possible unification given the state of the belief base.

- selectIntention(Queue<Intention> intentions): selects an intention to be further executed in the current reasoning cycle; it implements the S_I function in Figure 4.1. The default implementation removes and returns the first intention in the queue, and after execution the intention is inserted at the end of the queue, wich effectively means that the default implementation executes intentions in a 'round robin' like fashion (with only one plan body formula being executed in one round). If the set of intention that already started running, contains an atomic intention, this function is not called and the atomic intention is selected instead.

- socAcc(Message m): returns true if message m is socially acceptable. The default implementation returns true for all messages. In applications where security is an issue, this needs to be overridden because the agent would be susceptible even to simple denial of service attacks with the default socAcc() method.

- buf(List<Literal> percepts): updates the belief base with the given percepts and adds all changes that were actually carried out as new events in the set of events.

- brf(Literal add, Literal rem, Intention i): revises the belief base with a literal to be added (if any), a literal to be deleted (if any), and the Intention structure that required the belief change. This method is meant to be overridden in user classes, given that the default implementation does nothing. That is to say that there is no belief revision in *Jason* unless the user includes a particular implementation of a belief revision algorithm; we comment on a particular algorithm to be available with *Jason* soon in Section 11.2. Therefore, if a plan executes +b and another +~b, the belief base will be inconsistent. Unless the user is specifically interested in paraconsistency, some effort should be put into a brf() implementation (note that avoiding direct contradictions assuming the latest change is always correct is straightforward). This method returns two lists in an array, the first contains all actual additions to the belief base and the second all deletions. These lists are then used to generate the relevant events.

It is important to emphasise that the belief update function provided with *Jason* simply updates the belief base and generates the external events (i.e. additions and deletions of beliefs from the belief base) in accordance with current percepts. It does not guarantee belief consistency. As seen in Chapter 5, percepts are actually sent from the environment, as a list of literals stating everything that is true (and explicitly false too, unless closed-world assumption is used) in the environment. It is up to the developer of the environment model to make sure that contradictions do not appear in the percepts. More specifically, the default implementation of the belief update function works as follows, where B is the set of all literals currently in the belief base *which have a* source(percept) *annotation*[2], P is the received list of current percepts, and E is the set of events:

for all $b \in B$
   if $b \notin P$
      then delete b from B
         add $\langle -b, \top \rangle$ to E

for all $p \in P$
   if $p \notin B$
      then add p to B
         add $\langle +p[source(percept)], \top \rangle$ to E

As a first example of agent customisation, we will change the selection of events of the vacuum cleaner robot described in Chapter 5 to prioritise events created when dirt is perceived. The queue of events of this agent may have events created from perception (such as dirty, pos(r), and pos(l)) and internal events created by its intentions. The agent class that implements an $S_{\mathcal{E}}$ function that always selects dirty perception events first is shown below. The method selectEvent goes through all events in the queue and checks whether the trigger is equal to dirty[source(percept)]. If such an event exists, the function removes it from the queue and returns that event as the one to be handled in the current reasoning cycle. Otherwise, it calls the default implementation of the method (which, recall, returns the event at the beginning of the queue).

```
import jason.asSemantics.*;
import jason.asSyntax.*;
import java.util.*;

public class DirtyFocusAgent extends Agent {

    static Trigger focus = Trigger.parseTrigger(
                        "+dirty[source(percept)]");

    @Override
    public Event selectEvent(Queue<Event> events) {
```

---

[2]Recall that the source of information is annotated in the belief base, so we know exactly which beliefs originated from perception of the environment.

```
    Iterator<Event> i = events.iterator();
    while (i.hasNext()) {
        Event e = i.next();
        if (e.getTrigger().equals(focus)) {
          i.remove();
          return e;
        }
    }
    return super.selectEvent(events);
  }
}
```

This class is assigned to an agent in the project configuration file, as in the example below. Appendix A.2 shows all the features of the multi-agent system configuration file.

```
MAS vacuum_cleaning {
    environment: VCWorld
    agents:

        vc agentClass DirtyFocusAgent;
}
```

As a second example, suppose we want to indicate in the AgentSpeak code of some agent that an intention should be selected only when there is nothing else to be done; let us call this intention 'the idle intention'. More precisely, while the agent has many intentions (i.e. various concurrent foci of attention), the idle intention should never be selected. It will be selected only if it is the only intention of the agent. To mark an intention as 'idle', the programmer adds an idle annotation in the label of the plan that will form the intention. When this plan becomes an intended means of the intention, the whole intention becomes 'idle' (i.e. an intention to be executed only when the agent becomes otherwise idle). The code below implements an $\mathcal{S}_\mathcal{I}$ function that provides this functionality.

```
import jason.asSemantics.*;
import jason.asSyntax.*;
import java.util.*;

public class IdleAgent extends Agent {

  static Term idle = DefaultTerm.parse("idle");

  @Override
  public Intention selectIntention(Queue<Intention> intentions) {

    Iterator<Intention> ii = intentions.iterator();
```

```
        while (ii.hasNext()) {
          Intention i = ii.next();
          if (isIdle(i)) {
            if (intentions.size() == 1) {
              // there is only one intention and it has
              // the "idle" annotation, returns that one
              ii.remove();
              return i;
            }
          } else {
            // the current intention is not idle,
            // so it has higher priority
            ii.remove();
            return i;
          }
        }
        return null;
      }

      /** returns true if the intention has a "idle" annotation */
      private boolean isIdle(Intention i) {
        // looks for an "idle" annotation in every
        // intended means of the intention stack
        for (IntendedMeans im : i.getIMs()) {
          Pred label = im.getPlan().getLabel();
          if (label.hasAnnot(idle)) {
            return true;
          }
        }
        return false;
      }
    }
```

To exemplify this class in action, consider an agent with following code:

```
    /* Initial goals */

    !start(5).
    !free(5).

    /* Plans */

    +!start(0).
    +!start(X) <- .print(X); !start(X-1).

    +!free(0).
    +!free(X) <- .print(free); !free(X-1).
```

Its execution output is:

```
[a] saying: 5
[a] saying: free
[a] saying: 4
[a] saying: free
[a] saying: 3
[a] saying: free
[a] saying: 2
[a] saying: free
[a] saying: 1
[a] saying: free
```

Note that the two intentions are executed concurrently. If we add a label annotated with idle in the last plan

```
@l[idle] +!free(X) <- .print(free); !free(X-1).
```

the execution is as expected:

```
[a] saying: 5
[a] saying: 4
[a] saying: 3
[a] saying: 2
[a] saying: 1
[a] saying: free
[a] saying: free
[a] saying: free
[a] saying: free
[a] saying: free
```

As shown in this example, it is possible to extend the *Jason* programming language to add new features such as the 'idle' intention. However, it should be emphasised that tweaking with the selection functions can be dangerous, as the typical problems of concurrent programming can happen when new selection functions are introduced without due care, for example starvation of intentions or events. On the other hand, practical agents might need to starve events if the environment is too dynamic anyway.

Advance

Regarding customisation of $S_\mathcal{O}$, it is also reasonable to give a word of caution for programmers. As in Prolog, programmers often take advantage of the default style of option selection, for example by writing the plan for the end of a recursive plan first. However, some applications might benefit from meta-level information that would allow, e.g., for decision-theoretic selection of a plan, and in that case an agent-specific option selection function needs to be implemented. Therefore care should be taken by the programmer, so that the customised $\mathcal{S}_O$ function either makes sure the end of recursion is selected whenever appropriate or that it still uses the order in which plans are written for selecting among applicable plans that use recursion.

7.3 Customising the Overall Architecture

For an agent to work effectively in a multi-agent system, it must interact with the environment and other agents; the AgentSpeak interpreter is only the reasoning module within an 'overall agent architecture' that interfaces it with the outside world. We call it this way to avoid confusion with the fact that BDI is the agent architecture, but this is just the cognitive part of the agent, so to speak. In this section we use the term 'architecture' to mean the overall agent architecture. Such architecture provides perception (which models the agent's 'sensors'), acting (modelling the agent's 'effectors'), and how the agent receives messages from other agents. These aspects can also be customised for each agent individually.

The default implementation of the architecture is in the AgArch class that the user can extend if necessary (see the class diagram in Figure 7.3). This class is a 'bridge' to the underlying multi-agent infrastructure[3] so that the programmer can customise it without caring about the concrete implementation of perception and communication done by the infrastructure. The methods normally overridden are:

- perceive(): this method should return a list of literals that represent what has just been perceived by the agent. The default implementation simply gets the perceptions sent by the environment and returns it. This method can be customised, for example, to change the agent's perception capabilities (e.g. to test the agent's performance under faulty perception).

- act(ActionExec action, List<ActionExec> feedback): when the execution of some intention contains an action, this method is called to perform the action in the environment, where the first argument contains details of the action. In simulated environments, this means to simply send a request message for the environment implementation to simulate the action execution (this is the default implementation). In agents that have concrete effectors, this method should interact with the hardware to execute the action. While the action is being performed, the intention that made the action execution choice is suspended. When the action is finished (successfully or not), it

---

[3]The actual infrastructure of a MAS is set up in the configuration project with the `infrastructure` entry. Possible values are `Centralised`, `Saci` and `JADE` [8], but others may be available in future or even be implemented by the user.

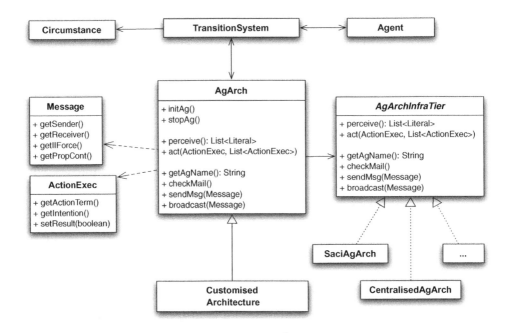

Figure 7.3 Agent architecture class diagram.

should be added in the feedback list so that the intention can be resumed or failed.

- sendMsg(Message m): this method sends a message to another agent. The default implementation simply calls the infrastructure method which effectively does that.

- broadcast(Message m): this method sends a message to all known agents sharing the environment (i.e. the agent society). The default implementation also just calls the infrastructure method.

- checkMail(): this method reads all messages in the agent's mailbox located at the infrastructure level (where messages could be in KQML, FIPA-ACL or some other format) and adds them in the mailbox at the interpreter level as instances of the *Jason*'s Message class. Again, the default implementation calls the infrastructure method to perform this task.

When customising some of these methods, it is often appropriate to use the default implementation defined in AgArch and, after it has done its job, process the return of those methods further. For example, to remove all positioning-related perception for the vacuum cleaning agent, the following class can be used for the agent's architecture:

```java
import jason.asSyntax.*;
import jason.architecture.*;
import java.util.*;

public class VCArch extends AgArch {

  @Override
  public List<Literal> perceive() {

     // gets the default perception
     List<Literal> per = super.perceive();

     // perception is null when nothing
     // has changed in the environment
     if (per != null) {
        Iterator<Literal> ip = per.iterator();
        while (ip.hasNext()) {
          Literal l = ip.next();
          if (l.getFunctor().equals("pos")) {
             ip.remove();
          }
        }
     }
     return per;
  }
}
```

In a similar way to agent customisation, this architecture class is assigned to an agent in the project configuration file using the `agentArchClass` keyword, for example:

```
MAS vacuum_cleaning {
   environment: VCWorld
   agents:

       vc agentArchClass VCArch;
}
```

As a second example, suppose we want to control, at the architectural level, the vacuum cleaner actions so that only the actions `left` and `right` are actually performed. All other actions the robot may choose to do will not be executed at all, although the interpreter should go ahead as though they were successfully executed. This kind of architecture customisation may be used, for instance, to prevent certain agents to try to do something dangerous or forbidden, given that they are autonomous and therefore it is difficult to ensure that they will not attempt to do certain things. The method act below controls this for the vacuum cleaning example:

```
@Override
public void act(ActionExec action, List<ActionExec> feedback) {
    String afunctor = action.getActionTerm().getFunctor();
    if (afunctor.equals("left") || afunctor.equals("right")) {

        // pretend that the action was successfully performed
        action.setResult(true);
        feedback.add(action);
    } else {

        // calls the default implementation
        super.act(action,feedback);
    }
}
```

As a last example of architecture customisation, we return to the domestic robot example introduced in Section 3.4. Suppose that the supermarket agent should ignore messages sent by owner agents. We can easily change the method check-Mail() to remove such messages at the architectural level (see also exercise 4 for an alternative and more usual way of eliminating unacceptable messages using agent rather than architecture customisation). The purpose here is simply to illustrate how the checkMail method can be customised. A more typical use of checkMail() customisation would be, for example, to convert messages into a symbolic format in case AgentSpeak agents are interacting with heterogeneous agents.

```
@Override
public void checkMail() {

    // calls the default implementation to move all
    // messages to the circumstance's mailbox.
    super.checkMail();

    // gets an iterator for the circumstance mailbox
    // and removes messages sent by agent "owner"
    Iterator im = getTS().getC().getMailBox().iterator();
    while (im.hasNext()) {
        Message m = (Message) im.next();
```

```
        if (m.getSender().equals("owner")) {
            im.remove();

            // sends a message to owner to inform that
            // the received message was ignored
            Message r = new Message(
                "tell",
                getAgName(),
                m.getSender(),
                "msg(\"You are not allowed to order anything, "+
                    "only your robot can do that!\")"
            );

            sendMsg(r);
        }
    }
}
```

Note that this behaviour can also be programmed at the AgentSpeak level, writing a plan to ignore all messages from the owner:

```
+X[source(owner)] <- -X.
```

However it is conceptually very different to decide not to process these messages at the cognitive level (in AgentSpeak) – where it is a *decision* of an autonomous agent whose practical reasoning is declaratively represented – rather than simply never knowing that some message was sent in the first place. This brings up the question of what is the right abstraction level in *Jason* to program a given system requirement. We can program almost everything in Java as user-defined components, but in many cases we lose the higher abstraction level and flexibility of AgentSpeak. On the other hand, some aspects are easier to program in Java yet without compromise to the representation of the agent's practical reasoning. There is no general answer, since each MAS project (or application domain) will have very different requirements and design characteristics. Nevertheless, developers should be aware of the issue so as to make reasoned design and implementation decisions rather than being extreme in either using too much or too little customisation code. We say more on this issue in Section 11.3.

7.4 Customising the Belief Base

The agent's belief base (BB) itself can be customised, which can be quite useful, for example in large-scale applications. The default implementation is in the

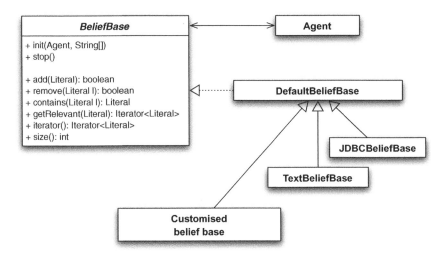

Figure 7.4 Belief base class diagram.

DefaultBeliefBase class, which implements the methods of the interface BeliefBase (see Figure 7.4). The main methods of this interface are:

- add(Literal bel): adds the literal in the belief base. It returns true if the belief was added. If the belief is already there, it returns false. If the belief given as a parameter is in the BB except for (some of) the annotations, only those annotations are included in the belief that already is in the BB.

- remove(Literal bel): removes the literal from the belief base. It returns true if the belief was actually removed. Again it might be the case that only some annotations need to be removed.

- contains(Literal bel): checks whether the belief is in the BB. It returns the literal as it is in the BB, and annotations in the parameter should be ignored by this method. For instance, if the BB is {a(10) [a,b]} and the parameter is a(10), the returned literal is a(10) [a,b]. If the BB does not contain the belief, it returns null.

- getRelevant(Literal bel): returns an iterator for all beliefs with the same functor and arity as the parameter belief. For instance, if the BB is {a(10),a(20), a(2,1), b(f)} and the parameter is a(5), the return of this method is an iterator for the list {a(10),a(20)}.

There are two BB customisations available with the *Jason* distribution: one which stores the beliefs in a text file (so as to persist the state of an agent's belief base) and another which stores some of the beliefs in a relational database. This latter customisation can be used the access any relational database. For example, a very simple agent that wants to count how many times it has been executed may use the following code:

```
!count. // initial goal.

+!count : not count_exec(_) <-  +count_exec(1).
+!count :     count_exec(X) <- -+count_exec(X+1).
```

Its execution with the default BB, that does not persist the beliefs, will always select the first plan. To configure agent a to use the BB that persists all beliefs in a single text file, the declaration of the agent in the project configuration is:

```
MAS custBB {
    agents:
        a beliefBaseClass jason.bb.TextPersistentBB;
}
```

Now every time this agent is executed, its beliefs are loaded from a file (called <agent name>.bb) and, before finishing, its beliefs are stored in the same file. Note that the AgentSpeak code remains the same regardless of the belief base customisation. The implementation of this BB simply overrides the init and stop methods to load and save the beliefs respectively.

To use persistence in a database, the Java JDBC (Java DataBase Connectivity) feature is used. The connection to the database requires some parameters, so five parameters should be informed to the BB customisation class: the JDBC driver, the connection URL for the database, the user, the password, and a list of the beliefs that are mapped into tables of the database. Each item in the list is formed by a structure where the functor is the belief to be stored, the first argument is its arity and the second is the table name when it is to be stored. Where this second argument is omitted, the functor is used as the table name. This customisation thus does not store all beliefs in a database, but only those explicitly listed. For example, to store the belief count_exec, that has arity 1, in a table named tablece the agent declaration is:

```
MAS custBB {
    agents:

        a beliefBaseClass jason.bb.JDBCPersistentBB(
            "org.hsqldb.jdbcDriver", // driver for HSQLDB
            "jdbc:hsqldb:bookstore", // URL connection
            "sa", // user
            "",   // password
            "[count_exec(1,tablece)]");
}
```

If the table `tablece` does not exist in the database, it will be created. With this BB configuration, all consulting or changes to the `count_exec` belief are mapped to SQL commands by the BB customisation.

This BB can also be used to access tables from legacy systems. Suppose there is a database with a table called `author_table` with three columns: author identification, name and e-mail. To access this table, the entry in the list of tables should be `author(3,author_table)`. In the AgentSpeak code, this table can be consulted or changed as an ordinary belief:

```
+someevent
    : author(Id, "Morgana", Email) // consult Morgana's e-mail
    <- -+author(Id,"Morgana","x@acme.com")... // and updates it

+otherevent
    <- .findall(Id, author(Id,_,_),L);
       .print("All Ids = ", L).
```

The last example we give in this section is the Java code for a BB customisation that annotates every belief with the time it was added to the BB, where the time stamp is given in milliseconds since the multi-agent system started running. The code essentially overrides the method **add** to include the annotation in beliefs that do not already have such annotation.

```
import jason.asSemantics.*;
import jason.asSyntax.*;
import jason.bb.*;
import java.util.*;

public class TimeBB extends DefaultBeliefBase {

  private long start;

  @Override
  public void init(Agent ag, String[] args) {
     start = System.currentTimeMillis();
     super.init(ag,args);
  }

  @Override
  public boolean add(Literal bel) {
     if (! hasTimeAnnot(bel)) {
        Structure time = new Structure("add_time");
        long pass = System.currentTimeMillis() - start;
        time.addTerm(new NumberTermImpl(pass));
        bel.addAnnot(time);
```

```
            }
            return super.add(bel);
        }

    private boolean hasTimeAnnot(Literal bel) {
        Literal belInBB = contains(bel);
        if (belInBB != null)
            for (Term a : belInBB.getAnnots())
                if (a.isStructure())
                    if (((Structure)a).getFunctor().equals("add_time"))
                        return true;
        return false;
    }
}
```

Considering the test code:

```
!start.
+!start
    <- +bel(1);
        +bel(1);
        +bel(2);
        myp.list_bels.
```

and the internal action `myp.list_bels` which lists the contents of the BB:

```
package myp;

import jason.asSemantics.*;
import jason.asSyntax.*;

public class list_bels extends DefaultInternalAction {

    @Override
    public Object execute(TransitionSystem ts,
                          Unifier un,
                          Term[] args) throws Exception {

        for (Literal b: ts.getAg().getBB()) {
            ts.getLogger().info(b.toString());
        }
        return true;
    }
}
```

the result of the execution is:

```
bel(1)[source(self),add_time(71)]
bel(2)[source(self),add_time(91)]
```

7.5 Pre-Processing Directives

As it is common in other languages, the AgentSpeak language as interpreted by
Jason accepts compiler directives. Directives are used to pass some instructions to
the interpreter that are not related to the language semantics, but are merely syntactical. Two kinds of directives may be inserted at any point in some AgentSpeak
code: single-command directives and scope directives.

Single-command directives have the following syntax:

```
"{" <directive-name> "(" <parameters> ")" "}"
```

There is one pre-defined single-command directive in *Jason* that is used to include
the AgentSpeak code in another file. The compiler loads the file, parses it, and adds
all beliefs, rules, initial goals, and plans into the agent. The code below exemplifies
this directive.

```
...
{ include("c.asl") }

bel(1).
+te : true <- action.
...
```

Scope directives have inner plans and beliefs; the syntax is as follows:

```
"{" "begin" <directive-name> "(" <parameters> ")" "}"
  <agentspeak-program>
"{" "end" "}"
```

This directive is usually used to change the inner plans (the plan patterns for declarative goals described in Chapter 8 are implemented using this kind of directives).
For example, suppose that for some reason we need to add (and easily remove)
.print(<X>) (where <X> is some parameter) at the end of each plan, for a large
number of plans. We can create a new directive, called, say, add_print, and use it
as in the following program:

```
{ begin add_print("end of plan") }

+te : b1 <- act1.
+te : b2 <- act2.

< all other plans >

{ end }
```

The `add_print` directive would change the inner plans to:

```
+te : b1 <- act1; .print("end of plan").
+te : b2 <- act2; .print("end of plan").

< all other plans also changed >
```

To stop adding the print in the end of the plans, we simply comment out the directive.

An interesting features of *Jason* is that the set of compiler directives can be effortlessly extended by the programmer. Each new directive is programmed in a class that implements the interface Directive. This interface has only one method:

- Agent process(Pred directive, Agent outterContent, Agent innerContent): the parameter directive (a predicate) represents the directive and its parameters (e.g. `add_print("end plan")`), outterContent points to the agent where this directive is being used, and innerContent points to the content enclosed by the directive. The inner content has the set of plans and beliefs to be modified by the directive. The method should return a new instance of agent where the modified plans and beliefs are added. Those new beliefs and plans will be then included in the agent that calls the directive.

The following Java class implements the `add_print` directive:

```java
package myp;

import jason.asSyntax.*;
import jason.asSyntax.directives.*;
import jason.asSyntax.BodyLiteral.BodyType;

public class AddPrint implements Directive {

  public Agent process(Pred directive,
                       Agent outterContent,
                       Agent innerContent) {

    Agent newAg = new Agent();
    Term arg = directive.getTerm(0); // get the parameter
    for (Plan p: innerContent.getPL()) {
      // create the new command .print(X) -- a body literal
      Literal print = new Literal(".print");
      print.addTerm((Term)arg.clone());
      BodyLiteral bl =
            new BodyLiteral(BodyType.internalAction, print);

      p.getBody().add(bl); // appends the new formula to the plan
```

```
        newAg.getPL().add(p);   // includes the plan in the PL
    }
    return newAg;
  }
}
```

To inform the compiler that this new directive is available, the keyword `directives` should be used in the project configuration file:

```
MAS ... {
    agents: ...

    directives: add_print = myp.AddPrint;
}
```

7.6   Exercises

1. Write an internal action that computes the square root of the number given [Basic] as parameter.

2. For an agent using the customised belief base where beliefs have their cre- [Advanc] ation time as annotations, write an internal action that removes all beliefs in the belief base that are older than some value informed as argument.

3. For the same agent as the previous exercise, write an internal action that [Advanc] returns all beliefs in the belief base that are older than some value informed as argument. This internal action should backtrack for all possible values. This is an example of the internal action usage:

```
+someevent
    :   myp.older_bel(b(X),1000) & // if b(X) is older than 1 sec.
        X > 10                     //    and X > 10
    <- ...
```

4. In the example of the supermarket agent architecture, used to avoid messages [Basic] from owner agents, rather than customising the method checkMail in the architecture, customise the method socAcc in an agent class customisation class instead.

5. Create a new special annotation, identified by `priority(p)` (where p is an [Advanc] integer number), that defines priorities for intention selection. The intention selection function should select intentions according to the priority: greater values of p denote grater priority. It should work so that only when intentions with higher priority have finished (or changed their priority) are

intentions with lower priority selected. The default priority is 0. Depending on the state of the stack of intended means, there could be various different ways of determining the priority of a given intention:

```
!g1.

+!g1[priority(-10)]
   :  true    // the priority of an intention where this plan
   <- !g2.    // becomes the topmost intended means is -10 ...

+!g2[priority(20)]
   :  true    // ... but changes to priority 20 instead when
   <- ...     // this plan becomes the topmost intended means
```

For this exercise, consider that the priority of the highest intended means within the intention should be used to select the intention.

anced 6. Implement an agent architecture that uses the Java RMI (Remote Method Invocation) API to send and receive messages.

Basic 7. Write a customised belief base for an 'amnesiac' agent whereby nothing is ever added to the belief base. In other words, the agent immediately forgets everything it attempts to store in the belief base.

Basic 8. Implement a belief base with a graphical interface that shows the current beliefs of the agent.

anced 9. Extended the previous exercise for an application where it is known that the beliefs in the belief base form a particular structure, say, a graph. The graphical interface should show the graph rather than a sequence of literals in textual form.

anced 10. Implement a belief base that supports a special type of belief, called 'interval belief'. When a belief such as `interval(a(1),a(400))` is in the BB, consulting any `a(i)` (where $1 \le i \le 400$) should successfully return `a(i)[interval]`.

anced 11. Create a new directive so that, for each sent message, a copy of the message is also sent to a list of agents. For instance, the code:

```
{ begin send_copy([monitor,bob]) }
+te : true
   <- .send(A,tell,vl(10));
      action.
{ end }
```

should be rewritten by the directive to:

```
+te : true
   <- .send(A,tell,vl(10));
      .send([monitor,bob],tell,sent(A,tell,vl(10)));
      action.
```

If the message copy is required for every single message sent by the agent, think of two other ways in which you could do this in *Jason*. Another useful thing to do is to keep copies in the belief base so that the agent can 'remember' (some of) the messages itself has sent.

8

Advanced Goal-Based Programming

In this chapter, we focus on one aspect of *Jason* programming style: the use of *goals*. One of the most important aspects of programming multi-agent systems is in the *use* of goals. The simplest type of goal is an *achievement goal*: where we want an agent to achieve some particular state of affairs ('win the game'). *Maintenance goals* are similarly basic: with such goals, we want our agent to maintain some state of affairs ('ensure the reactor temperature stays below 100 degrees'). However, *combinations* of achievement and maintenance goals are also common: for example, we might want to maintain some state of affairs until some particular condition becomes true. AgentSpeak and *Jason* provide direct support for achievement goals: this is the basic '$!\phi$' subgoal construct of plans. However, richer goal structures are not directly supported. There has been much discussion of such *declarative goals* in the agent programming literature [23, 55, 83, 96, 98, 102], as well as the different types of commitments that agents can exhibit towards their goals. This is essential in giving agents what can be called *rational behaviour*, in the sense that we do not expect agents to give up achieving their goal unless it has become impossible to achieve or the motivation for the goal no longer exists, in which case we do not expect agents to pointlessly try to achieve their goal. We believe that providing explicit support for complex goal patterns of this kind will be a major advantage when developing many multi-agent applications.

In *Jason*, we have avoided changing the language to explicitly support more complex goal structures. Instead, we use *pre-processing directives*, which transform plans according to certain patterns which then result in the goal behaviour of interest. The patterns as presented below were first introduced in [55]; we later show how the pre-processing directives can be used for each of the plan patterns. Before we explain the plan patterns, we will comment on how the BDI notions map into *Jason* programming and some internal actions available in *Jason* to allow for what could be called 'BDI programming' or 'goal-based programming'.

8.1   BDI Programming

In Chapter 2, we saw that an idealised BDI agent would deliberate and then use means-ends reasoning to determine the best course of action to achieve its current goals. In practice, however, this is often too computationally expensive to be feasible for large applications. In Chapter 3, we saw that AgentSpeak uses the intuitive notions of the BDI architecture to provide high-level programming constructs, and let programmers do much of this job. However, it helps to have a clear picture of how the BDI notions map into programming constructs and data structures of the interpreter in *Jason*. In Chapter 10 we give a more formal account of this relation. Before we start, it is worth recalling that the notions of *desire* and *goal* are very closely related. One normally uses 'goals' to refer to a consistent subset of an agent's 'desires'; in any case, both refer to states of affairs that the agent would like to bring about.

The notion of belief is straightforward as an agent in *Jason* has a belief base which contains a number of literals denoting the things directly believed by the agent. Intentions have two well-known aspects: they can be defined as the chosen course of action to achieve a goal (intention-to) or as the state that the chosen course of action aims to achieve (intention-that), in which case intentions are seen as a subset of the agent's desires, the ones it committed to bring about. The notion of intention-to is very clear in an AgentSpeak agent's intentional structure (see the 'set of intentions' in Figure 4.1): any intended means in that structure (i.e. an instance of a plan) is an intention-to. The intention-that notion, in particular in the declarative agent programming style that we discuss in this chapter, is naturally seen as the goal in the triggering event of a plan instance within the set of intentions. For example, if an agent has a plan '+!g : b <- a.' in its set of intentions, then a is an intention-to because g is an intention-that: the agent has the desire to achieve a state of affairs where g is true, and is committed to bring that about by being committed to executing a.

The notion of desire is a bit trickier. We just saw that intentions-that are a subset of desires. In particular, they are the specific desires for which the agent has chosen a course of action (and committed to execute it). These are the goals for which a plan is already in the set of intentions. Recall that, when new goals are created, they generate (internal) events that are included in the set of events (see Figure 4.1 again). At that point, the agent has a new goal (i.e. a new desire) but has not yet committed to bringing it about; in fact, it may turn out that the agent might not have the know-how, the ability, or the resources to do so. While a relevant and applicable plan is not found for a particular goal currently in the set of intentions, we can refer to it as a *desire*, but it is certainly not an intention.

While in AgentSpeak the context part of a plan implicitly checks the agent's belief state, it does not allow for the agent's desires and intentions to be checked. Also, another fundamental part of a BDI agent is to *drop* goals, which is also not part of the AgentSpeak language. We have addressed these shortcomings through the use of some standard internal actions, as follows. Recall that standard internal actions, as opposed to user-defined internal actions, are those available with the *Jason* distribution; they are denoted by an action name starting with the '.' symbol. (We list all standard internal actions in Appendix A.3.)

> **.desire:** this internal action takes a literal as parameter and succeeds if that literal is one of the agent's desires. In our framework, desires are achievement goals that appear in the set of events (i.e. internal events) or appear in intentions, including suspended ones. More specifically, the .desire action succeeds if the literal given as a parameter unifies with a literal that appears in a triggering event that has the form of a goal addition (i.e. +!g) within an event in the set of events or in the triggering event part of the head of any of the intended means (plans) in any of the agent's current intentions. One of the common views is that an intention is a desire (goal) which the agent is committed to achieving (by executing a plan, i.e. a course of action); goals in the set of events are those for which no applicable plan has become intended yet. Formal definitions are given in Section 10.4.

> **.intend:** similar to .desire, but for an intention specifically (i.e. excluding the achievement goals that are internal events presently in the set of events).

> **.drop_desire:** receives a literal as parameter and removes events that are goal additions with a literal that unifies with the one given as parameter, then does all that .drop_intention (described next) would do. That is, all intentions and events which would cause .desire to succeed for this argument are simply dropped.

> **.drop_intention:** as .drop_desire, but works on each intention in the set of intentions where any of the goal additions in the triggering events of plans unify with the literal given as parameter. In other words, all intentions which would make .intend true are dropped.

.drop_event: this is complementary to the two internal actions above, to be used where only achievement goals appearing in events (desires that are not intentions) should be dropped.

.drop_all_desires: no parameters needed; all events and all intentions are dropped; this also includes external events (differently from the actions above).

.drop_all_intentions: no parameters needed; all intentions except the currently selected intention (the one where .drop_all_ intentions appears) are dropped.

.drop_all_events: no parameters needed; all events (including external events) currently in the set of events are dropped.

We now take a closer look at two very important standard internal actions. They are particularly useful in combination with the plan failure handling mechanism introduced in Section 4.2, and are used in the plan patterns we introduce later in this chapter.

The internal actions often used in combination with plan failure are .succeed_goal and .fail_goal. The first, .succeed_goal(g), is used when the agent realises the goal has already been achieved, so whatever plan was being executed to achieve that goal no longer needs to be executed. The second, .fail_goal(g), is used when the agent realises that the goal has become impossible to achieve, therefore the plan that required g being achieved as one of its subgoals has to fail. More specifically, when .succeed_goal(g) is executed, any intention that has the goal g in the triggering event of any of its plans will be changed as follows. The plan with triggering event +!g is removed and the plan below that in the stack of plans forming that intention carries on being executed, at the point after goal g appeared in the plan body. Goal g, as it appears in the .succeed_goal internal action, is used to further instantiate the plan where the goal that was terminated early appears. With .fail_goal(g), the plan for +!g is also removed, but an event for the deletion of the goal whose plan body required g is generated instead: as there is no way of achieving g, the plan requiring g to be achieved has effectively failed.

It is perhaps easier to see how these actions work with reference to Figure 8.1. The figure shows the consequence of each of these internal actions being executed. The plan where the internal action appeared is not shown; it is likely to be within another intention. Note that the state of the intentions, as shown in the figure, is not the immediate state resulting from the execution of either of the internal actions – i.e. not the state at the end of the reasoning cycle where the internal action was executed – but the most significant next state of the changed intention.

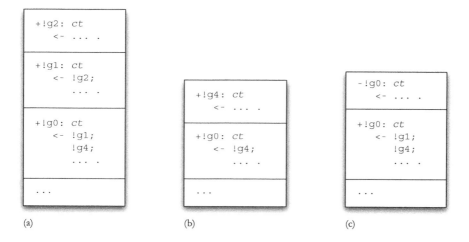

Figure 8.1 Standard internal actions for dropping goals. (a) Initial intention; (b) after `.succeed_goal(g1)`; (c) After `.fail_goal(g1)`.

8.2 Declarative (Achievement) Goal Patterns

Although goals form a central component of the AgentSpeak conceptual framework, it is important to note that the language itself does not provide any explicit constructs for handling goals with complex structures. As we noted above, a programmer will often think in terms of goals such as 'maintain P until Q becomes true', or 'prevent P from becoming true'. Creating AgentSpeak code to realise such complex goals can often be an essentially *ad hoc* process, dependent upon the experience of the programmer. A recent extension of *Jason* defined a number of declarative goal structures which can be realised in terms of *patterns* of AgentSpeak plans – that is, complex combinations of plan structures which are often useful in actual scenarios. Such patterns can be used to implement, in a systematic way, not only complex types of declarative goals, but also the types of agent commitments that they can represent, as discussed for example by Cohen and Levesque [29].

As an initial motivating example for declarative goals, consider a robot agent with the goal of being at some location (represented by the predicate `l(X,Y)`), and the following plan to achieve this goal:

```
+!l(X,Y) : bc(B) & B >0.2 <- go(X,Y).
```

where the predicate `bc/1` stands for 'battery charge', and `go` identifies an action that the robot is able to perform in the environment.

At times, using an AgentSpeak plan as a procedure can be quite useful as a programming practice. Thus, in a way, it is important that the AgentSpeak interpreter does not enforce any declarative semantics to its only (syntactically defined) goal construct. However, in the plan above, $l(X, Y)$ is clearly meant as a declarative goal; that is, the programmer expects the robot to believe $l(X, Y)$ (by perceiving the environment) if the plan executes to completion. If it fails because, say, the environment is dynamic, the goal cannot be considered achieved and, normally, should be attempted again.

*As much as possible, **Jason** programmers should bear in mind, when writing their code, that this is usually the right approach to thinking about goals that agents should achieve. While we find the freedom to use a goal simply as a procedure useful, it is by the declarative use of goals that we can determine whether some AgentSpeak code is using an appropriate style for agent programming. If you cannot find clearly declarative use of goals in your code, either your design is not appropriate, or your programming style is not right, or your application is not suitable for multi-agent systems.*

The type of situation in the example above is commonplace in multi-agent systems, and this is why it is important to be able to define declarative goals in agent-oriented programming. As pointed out by van Riemsdijk *et al.* [98], one can easily transform the above procedural goal into a declarative goal by adding a corresponding *test goal* at the end of the plan's body, as follows:

```
+!l(X,Y) : bc(B) & B >0.2 <- go(X,Y); ?l(X,Y).
```

This plan only succeeds if the goal is actually (believed to be) achieved; if the given (procedural) plan executes to completion (i.e. without failing) but the goal happens not to be achieved, the test goal at the end will fail. In this way, we have taken a simple *procedural* goal and transformed it into a *declarative* goal – the goal to achieve some state of affairs. We will see later that the plan failure mechanism in *Jason* can be used to account for the various attitudes that agents can have due to a declarative goal not being achieved (e.g. because the test goal at the end of the plan failed).

The above solution for simple declarative goals can be thought of as a *plan pattern*, which can be applied to solve other similar problems. This approach is inspired by the successful adoption of *design patterns* in object-oriented design [48].

To represent such patterns with AgentSpeak, we make use of skeleton programs with meta variables. For example, the general form of an AgentSpeak plan for a simple declarative goal, such as the one used in the robot's location goal above, is as follows:

```
+!g : c <- p; ?g.
```

Here, *g* is a meta variable that represents the declarative goal, *c* is a meta variable that represents the context expression stating in which circumstances the plan is applicable, and p represents the procedural part of the plan body (i.e. a course of action to achieve g). Note that, with the introduction of the final test goal, this plan to achieve g finishes successfully only if the agent believes g after the execution of plan body p.

To simplify the presentation of the patterns, we also define pattern rules which rewrite a set of AgentSpeak plans into a new set of AgentSpeak plans according to a given pattern. The following pattern rule, called **DG** (declarative goal), is used to transform procedural goals into declarative goals. The pattern rule name is followed by the (subscript) parameters which need to be provided by the programmer, besides the actual code (i.e. a set of plans) on which the pattern will be applied.

```
+!g : c1 <- p1.
+!g : c2 <- p2.
...
+!g : cn <- pn.
```
————————————————————————— $\textbf{DG}_g$ $(n \geq 1)$
```
+!g : g <- true.
+!g : c1 <- p1; ?g.
+!g : c2 <- p2; ?g.
...
+!g : cn <- pn; ?g.
+g : true <- .succeed_goal(g).
```

Essentially, this rule adds ?g at the end of each plan in the given set of plans which have +!g as trigger event, and creates two extra plans (the first and the last plans above). The first plan checks whether the goal g has already been achieved – in such a case, there is nothing else to do. That last plan is triggered when the agent perceives that g has been achieved while it is executing any of the courses of action p_i $(1 \leq i \leq n)$ which aim at achieving g; in this circumstance, the plan being executed in order to achieve g can be immediately terminated. The internal action .succeed_goal(g) terminates such a plan with success (as explained in the previous section).

In this pattern, when one of the plans to achieve g fails, the agent gives up achieving the goal altogether. However it could be the case that, for such a goal, the agent should try another plan to achieve it, as in the 'backtracking' plan selection mechanism available in platforms such as JACK [101] and 3APL [35]. In those mechanisms, usually only when all available plans have been tried in turn and failed is the goal abandoned with failure, or left to be attempted again later on. The following rule, called **BDG** (backtracking declarative goal), defines this pattern based on a set of conventional AgentSpeak plans $\mathcal{P}$ transformed by the **DG** pattern (each plan in $\mathcal{P}$ is of the form +!g : c <- p.):

$$\mathcal{P}$$
$$\overline{\rule{8cm}{0pt}}\ \mathbf{BDG}_g$$
$$\mathbf{DG}_g\,(\mathcal{P})$$

```
-!g : true <- !!g.
```

The last plan of the pattern catches a failure event, caused when a plan from $\mathcal{P}$ fails, and then tries to achieve that same goal g again. Notice that it is possible that the same plan is selected and fails again, causing a loop if the plan contexts have not been carefully programmed. Therefore, the programmer would need to specify the plan contexts in such a way that a plan is only applicable if it has a chance of succeeding regardless of it having been tried already (recently).

Instead of worrying about defining contexts in such complex way, in some cases it may be useful for the programmer to apply the following pattern, called **EBDG** (exclusive backtracking declarative goal), which ensures that none of the given plans will be attempted twice before the goal is achieved:

```
+!g : c₁ <- b₁.
+!g : c₂ <- b₂.
...
+!g : cₙ <- bₙ.
```
$$\overline{\rule{9cm}{0pt}}\ \mathbf{EBDG}_g$$
```
+!g : g <- true.
+!g : not p(1,g) & c₁ <- +p(1,g); b₁.
+!g : not p(2,g) & c₂ <- +p(2,g); b₂.
...
+!g : not p(n,g) & cₙ <- +p(n,g); bₙ.
-!g : true <- !!g.
 +g : true <- .abolish(p(_,g)); .succeed_goal(g).
```

In this pattern, each plan i, when selected for execution, initially adds a belief[1] $p(i,g)$; goal g is used as an argument to p so as to avoid interference between various instances of the pattern for different goals. The belief is used as part of the plan contexts (note the use of not $p(i,g)$ in the contexts of the plans in the pattern above) to state that the plan should not be applicable in a second attempt (of that same plan within a single adoption of goal g for that agent).

8.3   Commitment Strategy Patterns

In the EBDG pattern, despite the various alternative plans, the agent can still end up dropping the intention with goal g unachieved, if all those plans become

---

[1]In the actual pattern implementation this predicate is slightly different to avoid name clash.

non-applicable. In BDI parlance, this is because the agent is not sufficiently *committed* to achieving the goal. We now introduce various patterns that can be used to program various forms of commitment strategies introduced in the BDI literature.

With a *blind commitment*, the agent can drop the goal only when it is achieved. This type of commitment towards the achievement of a declarative goal can thus be understood as *fanatical commitment* [81]. The $BC_{g,F}$ pattern below defines this type of commitment:

$$\frac{\mathcal{P}}{\mathbf{F}(\mathcal{P})}\;\mathbf{BC}_{g,F}$$

```
+!g : true <- !!g.
```

This pattern is based on another pattern rule, represented by the variable $\mathbf{F}$; $\mathbf{F}$ is often **BDG**, although the programmer can chose any other pattern (e.g. **EBDG** if a plan should not be attempted twice). Finally, the last plan makes the agent attempt to achieve the goal even in case where there is no applicable plan. It is assumed that the selection of plans is based on the order that the plans appear in the program and all events have equal chance of being chosen as the event to be handled in a reasoning cycle. If the selection functions have been customised, this pattern needs to be used with care or changed, unless the customisation takes into consideration the use of this pattern.

For most applications, **BC**-style fanatical commitment is too strong. For example, if a robot has the goal to be at some location, it is reasonable that it can drop this goal if its battery charge is getting very low; in other words, the agent has realised that it has become impossible to achieve the goal, so it is useless to keep attempting it. This is very similar to the idea of a persistent goal in the work of Cohen and Levesque: a persistent goal is a goal that is maintained as long as it is believed not achieved, but still believed possible [29]. In [102] and [23], the 'impossibility' condition is called 'drop condition'. The drop condition f (e.g. 'low battery charge') is used in the single-minded commitment (**SMC**) pattern to allow the agent to drop a goal if it becomes impossible:

$$\frac{\mathcal{P}}{\mathbf{BC}_{g,BDG}(\mathcal{P})}\;\mathbf{SMC}_{g,f}$$

```
+f : true <- .fail_goal(g).
```

This pattern extends the **BC** pattern adding the drop condition represented by the literal f in the last plan. If the agent comes to believe f, it can drop goal g, signalling failure (refer to the semantics of the internal action .fail_goal in the previous section). This effectively means that the plan in the intention where g appeared, which depended on g being achieved to then carry on the plan execution, must itself fail (as g is now impossible to achieve). However, there might be an alternative for that plan which does not depend on g, so that plan's failure handling may take care of such situation.

As we have a failure drop condition for a goal, we can also have a success drop condition, e.g., because the motivation to achieve the goal has ceased to exist. Suppose a robot has the goal of going to the fridge because its owner has asked it to fetch a beer from there; then, if the robot realises that its owner does not want a beer anymore, it should drop the goal [29]. The belief 'my owner wants beer' is the *motivation* (m) for the goal. The following pattern, called relativised commitment (**RC**) defines a goal that is relative to a motivation condition: the goal can be dropped with success if the agent no longer has the motivation for it.

$$\frac{\mathcal{P}}{\mathbf{BC}_{g,BDG}(\mathcal{P})} \mathbf{RC}_{g,m}$$

```
-m : true <- .succeed_goal(g).
```

Note that, in the particular combination of **RC** and **BC** above, if the attempt to achieve g ever terminates, it will always terminate with success, since the goal will be dropped only if either the agent believes it has been achieved (by **BC**) or m is removed from belief base.

Of course we can combine the last two patterns above to create a goal which can be dropped if it has been achieved, has become impossible to achieve, or the motivation to achieve it no longer exists, representing what is called an 'open-minded commitment'. The open-minded commitment pattern (**OMC**) defines this type of goal:

$$\frac{\mathcal{P}}{\mathbf{BC}_{g,BDG}(\mathcal{P})} \mathbf{OMC}_{g,f,m}$$

```
+f : true <- .fail_goal(g).
-m : true <- .succeed_goal(g).
```

For example, an impossibility condition could be 'no beer at location (X,Y)' (denoted below by ~b(X,Y)), and the motivation condition could be 'my owner wants a beer' (denoted below by wb). Consider the plan below as representing the single known course of action to achieve goal l(X,Y):

```
+!l(X,Y) : bc(B) & B > 0.2 <- go(X,Y).
```

When the pattern $\mathbf{OMC}_{l(X,Y),\sim b(X,Y),wb}$ is applied to the initial plan above, we get the following set of plans:

```
+!l(X,Y) : l(X,Y) <- true.
+!l(X,Y) : bc(B) & B > 0.2 <- go(X,Y); ?l(X,Y).
-!l(X,Y) : true <- !!l(X,Y).
+!l(X,Y) : true <- !!l(X,Y).
+~b(X,Y) : true <- .fail_goal(l(X,Y)).
-wb : true <- .succeed_goal(l(X,Y)).
```

8.4 Other Useful Patterns

Another important type of goal in agent-based systems are *maintenance goals*, whereby an agent needs to ensure that the state of the world will always be such that g holds. Such an agent will need plans to act on the events that indicate the maintenance goal may fail in the future. In realistic environments, however, agents will probably fail in preventing the maintenance goal from ever failing. Whenever the agent realises that g is no longer in its belief base (i.e. believed to be true), it will certainly attempt to bring about g again by having the respective (declarative) achievement goal. The pattern rule that defines a maintenance goal (**MG**), but particularly in the sense of realising the failure in a goal maintenance, is as follows:

$$\frac{\mathcal{P}}{\begin{array}{l} g[\text{source(percept)}]. \\ -g : \text{true} <- !g. \\ \mathbf{F}(\mathcal{P}) \end{array}} \mathbf{MG}_{g,F}$$

The first line of the pattern states that, initially (when the agent starts running) it will assume that g is true. (As soon as the interpreter obtains perception of the environment for the first time, the agent might already realise that such assumption was wrong.) The first plan is triggered when g is removed from the belief base, e.g. because g has not been perceived in the environment in a given reasoning cycle, and thus the maintenance goal g is no longer achieved. This plan then creates a declarative goal to achieve g. The type of commitment to achieving g if it happens not to be true is defined by $\mathbf{F}$, which would normally be **BC** given that the goal should not be dropped in any circumstances unless it is has been achieved again. (Realistically, plans for the agent to attempt pro-actively to prevent this from ever happening would also be required, but the pattern is useful to make sure the agent will act appropriately if things go wrong.)

We now show another useful pattern, called sequenced goal adoption (**SGA**). This pattern should be used when various instances of a plan to achieve a goal should not be adopted concurrently (e.g. a robot should not try to clean two different places at the same time, even if it has perceived dirt in both places, which will lead to the adoption of goals to clean both places). To solve this problem, the **SGA** pattern adopts the first occurrence of the goal and records the remaining occurrences as pending goals by adding them as special beliefs. When one such goal occurrence is achieved, if any other occurrence is pending, it gets activated.

—————————————————————————— $\text{SGA}_{t,c,g}$

```
t : not fl(_) & c <- !fg(g).
t : fl(_) & c <- +fl(g).
+!fg(g) : true <- +fl(g); !g; -fl(g).
-!fg(g) : true <- -fl(g).
-fl(_) : fl(g) <- !fg(g).
```

In this pattern, t is the trigger leading to the adoption of a goal g; c is the context for the goal adoption; $fl(g)$ is the flag to control whether an instance of goal g is already active; and $fg(g)$ is a procedural goal that guarantees that fl will be added to the belief base to record the fact that some instance of the goal has already been adopted, then adopts the goal $!g$, as well as guaranteeing that fl will be eventually removed whether $!g$ succeeds or not. The first plan is selected when g is not being pursued; it simply calls the fg goal. The second plan is used if some other instance of that goal has already been adopted. All it does is to remember that this goal g was not immediately adopted by adding $fl(g)$ to the belief base. The last plan makes sure that, whenever a goal adoption instance is finished (denoted by the deletion of an fl belief), if there are any pending goal instances to be adopted, they will be activated through the fg call.

8.5 Pre-Processing Directives for Plan Patterns

Once the above goal and commitment strategies type are well understood, it is very easy to implement them in *Jason* using the pre-processing directives for the various plan patterns. They all have a {begin ...} ... {end} structure. The plan-pattern directives currently available in *Jason* are as follows:

```
{begin dg(⟨goal⟩)}
    ⟨plans⟩
{end}
```

This is the 'declarative goal' pattern described above.

```
{begin bdg(⟨goal⟩)}
    ⟨plans⟩
{end}
```

This is the 'backtracking declarative goal' pattern described above. Recall that further patterns can be used in the body of the directive where the ⟨plans⟩ are given.

```
{begin ebdg(⟨goal⟩)}
    ⟨plans⟩
{end}
```

This is the 'exclusive backtracking declarative goal' pattern described above.

```
{begin bc(⟨goal⟩,⟨F⟩)}
    ⟨plans⟩
{end}
```

This is used to add a 'blind commitment' strategy to a given goal. F is the name of another pattern to be used for the goal itself, the one for which the commitment strategy will be added; it is typically **BDG**, but could be any other goal type.

```
{begin smc(⟨goal⟩,⟨f⟩)}
    ⟨plans⟩
{end}
```

This is the pattern to add a 'single-minded commitment' strategy to a given goal; f is the *failure condition*.

```
{begin rc(⟨goal⟩,⟨m⟩)}
    ⟨plans⟩
{end}
```

This is the pattern to add a 'relativised commitment' strategy to a given goal; m is the *motivation* for the goal.

```
{begin omc(⟨goal⟩,⟨f⟩,⟨m⟩)}
    ⟨plans⟩
{end}
```

This is the pattern to add an 'open-minded commitment' strategy to a given goal; f and m are as in the previous two patterns.

```
{begin mg(⟨goal⟩,⟨F⟩)}
    ⟨plans⟩
{end}
```

This is the 'maintenance goal' pattern; F is the name of another pattern to be used for the achievement goal, in case the maintenance goal fails. Recall this is a reactive form of a maintenance goal, rather than proactive. It causes the agent to act to return to a desired state after things have already gone wrong.

```
{begin sga(⟨t⟩,⟨c⟩,⟨goal⟩)}
    ⟨plans⟩
{end}
```

This is the 'sequenced goal adoption' pattern; the pattern prevents multiple instances of the same plan to be simultaneously adopted, where t is the triggering event and c is the context of the plan.

9

Case Studies

This chapter describes two *Jason* applications which are more elaborate than the examples presented in previous chapters. The purpose in presenting these applications is to demonstrate how *Jason* was used to develop solutions in those contexts.

The first application consists of a team of agents that took part in the second CLIMA Contest, held with CLIMA VII in 2006 [17, 33]. The main feature of this application is to integrate AgentSpeak and Java code, specially in order to use legacy code from within the agents and to interact with the contest simulation server. Even though the implementation was not completely finished in time for the contest, the team won the competition, and the experience also allowed us to improve various aspects of the interpreter. The version described here has some more improvements in the AgentSpeak code in comparison to the one used in the competition.

The second application is the implementation of the Electronic Bookstore designed as a case study in the 'Developing Intelligent Agent Systems' book by Padgham and Winikoff [75] using the Prometheus methodology. In the book, the system was implemented in JACK and we present here, in Section 9.2, its implementation using *Jason* instead. Although the main core of the application logic is developed in AgentSpeak, a Web-based interface, distributed agents and database access are also required in this application. Integration with these technologies can be useful for many applications of interest to readers of this book, therefore the description of this application here focuses on this integration, as the AgentSpeak code is rather simple. It is also useful for showing how *Jason* can be applied directly in the implementation of a system designed using a BDI-based agent-oriented software engineering methodology.

Programming Multi-Agent Systems in AgentSpeak using Jason   R.H. Bordini, J.F. Hübner, M. Wooldridge
© 2007 John Wiley & Sons, Ltd

9.1   Case Study I: Gold Miners

We quote below the general description of the CLIMA contest's scenario;
Figure 9.1 shows a screenshot.

> Recently, rumours about the discovery of gold scattered around deep
> Carpathian woods made their way into the public. Consequently
> hordes of gold miners are pouring into the area in the hope to collect
> as much of gold nuggets as possible. Two small teams of gold min-
> ers find themselves exploring the same area, avoiding trees and bushes
> and competing for the gold nuggets spread around the woods. The
> gold miners of each team coordinate their actions in order to collect as
> much gold as they can and to deliver it to the trading agent located in
> a depot where the gold is safely stored.
>
> (http://cig.in.tu-clausthal.de/CLIMAContest/)

To implement our team, two features of *Jason* were specially useful: architecture
customisation and internal actions (see Figure 9.2). A customisation of the overall
architecture, as explained in Section 7.3, is used to interface between the agent and
its environment. The environment for the CLIMA contest was implemented by

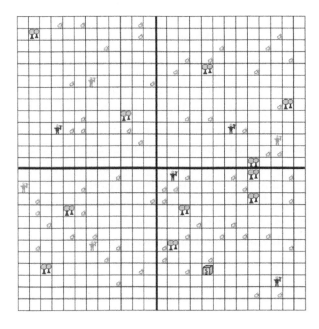

Figure 9.1 The contest scenario.

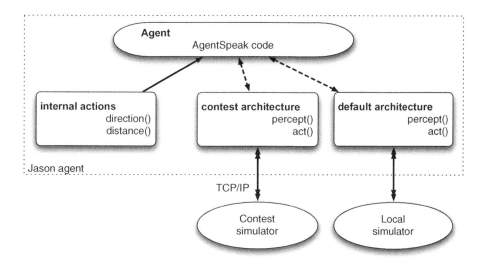

Figure 9.2 Agent extensibility and customisation.

the contest organisers in a remote server that simulates the mining field, sending perception to the agents and receiving requests for action execution according to a particular protocol also defined by the contest organisers. Therefore, when an agent attempts to perceive the environment, the architecture sends to it the information provided by the simulator, and when the agent chooses an action to be performed, the architecture sends the action execution request back to the simulator. For example, a plan such as:

```
+pos(X,Y) : Y > 0 <- left.
```

would be triggered when the agent perceives its position and its current line in the world grid is greater than zero. The +pos(X,Y) percept is produced by the architecture from the messages sent by the simulator, and left is an action that the architecture sends to the simulator for execution.

The architecture customisation was also very useful to bind the agent to an environment implementation, which could be either the real contest simulator or our own local simulator, by simply changing the project configuration. Since the contest simulator is not always available and the communication to/from it is quite slow, having a local simulator to develop and test the team was quite useful.

Although most of the agent is coded in AgentSpeak, some parts were implemented in Java, in this case because we wanted to use legacy code; in particular, we already had a Java implementation of the A* search algorithm, which we used to find paths in instances of the simulated scenario (it is interesting to note that in one of the maze-like scenarios used in the competition, our team was the only one to successfully find a path to the depot). This algorithm was made accessible to the agents by means of internal actions (refer to Chapter 7.1, which explains how new internal actions can be created). The code below exemplifies how these internal actions might be used in AgentSpeak plans.

```
+gold(X,Y)
    :   pos(X,Y) & depot(DX,DY) & carrying_gold
    <- pick;
        jia.get_direction(X,Y,DX,DY,Dir);
        Dir.
```

In this plan, when the agent perceives some gold in its current position, it picks up the gold and calls the `get_direction` internal action of the `jia` library. This internal action receives two locations as parameters ($\langle$X,Y$\rangle$ and $\langle$DX,DY$\rangle$), computes a path between them using A* (using the Manhattan distance as heuristic, as usual in scenarios such as this), and unifies `Dir` with the first action (up, down, left, or right) according to the path found from the first towards the second coordinate. The plan then asks for the execution of the action instantiated to variable `Dir`. Note that this plan is illustrative; it does not generate the behaviour of carrying the gold to the depot; only one step towards it is performed in the excerpt above. Since each agent has only a partial view of the environment and some kind of information is useful for all of them, especially the location of obstacles for A*, the customised architecture ensures that such information was shared among all agents of the *Jason* team.

Overall Team Strategy

The team is composed of two roles played by five agents. The miner role is played by four agents which will have the goal of finding pieces of gold and carrying them to the depot. The team also has one other agent enacting the leader role; its goals are to allocate agents to the quadrants closest to their initial position in the scenario and to allocate free agents to a piece of gold that has been found by any of the team members. Note that exactly four (mining) agents need to log in to the simulation server for each team. The diagrams in Figures 9.3 – 9.5 give an overview of the system and those two roles using the Prometheus methodology [75].

The overall strategy of the team is as follows. Each miner is responsible for systematically (rather than randomly) searching for gold *within* one quadrant of the environment (see Figure 9.6). Since the initial positions of the agents are only known when the game starts, the allocation of the agents to quadrants depends on such positions. The team uses the protocol in Figure 9.7 for quadrant allocation. At the beginning of each game, the four mining agents send their location to the

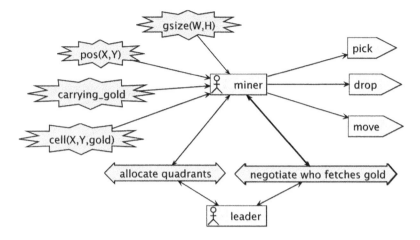

Figure 9.3 System overview diagram.

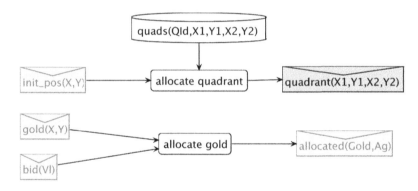

Figure 9.4 Leader agent overview diagram.

leader agent; the leader allocates each quadrant to an agent by checking which agent is nearest to that quadrant, and sends a message to each agent saying to which quadrant they have been allocated. We have decided to centralise some decisions in a leader agent so as to decrease the number of required messages in such distributed negotiations; even though all agents were run in the same machine in the actual competition, this is particularly important if in a future implementation we decide to run the agents in different machines, in case agents become too computationally heavy to run them all in one machine.

Another protocol is used to decide which agent will commit to retrieve a piece of gold found by a miner that is already carrying gold (Figure 9.8). When an agent (e.g. some miner in Figure 9.8) sees a piece of gold and cannot pick it up

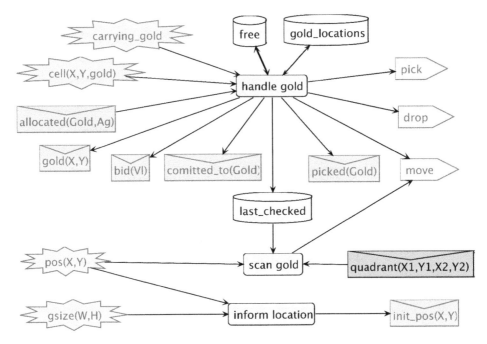

Figure 9.5 Miner agent overview diagram.

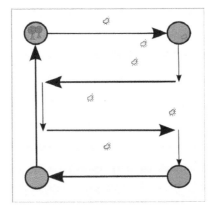

Figure 9.6 Miner's search path within its quadrant.

(e.g. because it is already carrying gold), it broadcasts the location of the piece of gold just found to all agents. The other agents bid to take on the task of fetching that piece of gold; such a bid is computed based on its availability and distance to that gold. All bids are sent to the leader who then chooses the best agent to commit to collect that piece of gold. This protocol also states that, whenever an agent

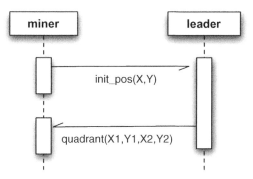

Figure 9.7 Quadrant allocation protocol.

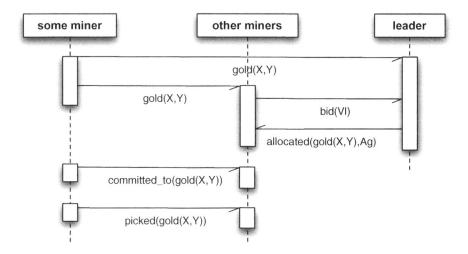

Figure 9.8 Gold allocation protocol.

decides to go towards some piece of gold or picks one, it should announce that fact to all other agents. Another agent that happens to also have the intention of fetching that same piece of gold can therefore drop the intention.

The miner agents have two mutually exclusive goals: 'find gold' and 'handle gold'. Whenever the agent has currently no other intention, it adopts the goal of exploring its own quadrant to find gold. When the agent either perceives some gold or is allocated to a piece of gold following the protocol in Figure 9.8, it gives up the 'find gold' goal and commits to the goal of handling that particular piece of gold. When this latter goal is achieved, the agent commits again to the 'find gold' goal.

Below, some important excerpts of the AgentSpeak code are presented; the complete code and the local simulator of the contest are available in the *Jason* distribution (in folder `examples/Gold-Miners`).

Implementation of Quadrant Allocation

An agent may participate in many simulations, each one with a particular scenario defining specific obstacles, grid size and depot location. When a new simulation starts, the architecture adds a belief with the simulation identification (represented by variable `S` in the sources) and the grid size of the scenario. Once they have this information, the miners wait to be informed of their initial location in that scenario and send this information to the leader:

```
+gsize(S,_,_) : true // S is the simulation round ID
  <- !send_init_pos(S).
+!send_init_pos(S) : pos(X,Y)
  <- .send(leader,tell,init_pos(S,X,Y)).
+!send_init_pos(S) : not pos(_,_) // if I do not know my position yet
  <- .wait("+pos(X,Y)", 500);        // wait for it and try again
     !!send_init_pos(S).
```

The leader allocates the miners quadrants based on their positions on the grid:

```
@quads[atomic]
+gsize(S,W,H)
  <- // calculate the area of each quadrant and remember them
     +quad(S,1, 0, 0, W div 2 - 1, H div 2 - 1);
     +quad(S,2, W div 2, 0, W-1, H div 2 - 1);
     +quad(S,3, 0, H div 2, W div 2 - 1, H - 1);
     +quad(S,4, W div 2, H div 2, W - 1, H - 1).

+init_pos(S,X,Y)[source(A)]
   : // if all miners have sent their position
     .count(init_pos(S,_,_),4)
  <- // remember who doesn't have a quadrant allocated
     // (initially all miners)
     +~quad(S,miner1); +~quad(S,miner2);
     +~quad(S,miner3); +~quad(S,miner4);
     !assign_all_quads(S,[1,2,3,4]).

+!assign_all_quads(_,[]).
+!assign_all_quads(S,[Q|T])
  <- !assign_quad(S,Q);
     !assign_all_quads(S,T).
```

```
// assign quadrant Q to a miner
+!assign_quad(S,Q)
   :  quad(S,Q,X1,Y1,X2,Y2) &
      ~quad(S,_) // there still is a miner without quadrant
   <- .findall(Ag, ~quad(S,Ag), LAgs);
      !calc_ag_dist(S,Q,LAgs,LD);
      .min(LD,d(Dist,Ag));
      .print(Ag, "'s Quadrant is: ",Q);
      -~quad(S,Ag);
      .send(Ag,tell,quadrant(X1,Y1,X2,Y2)).

+!calc_ag_dist(S,Q,[],[]).
+!calc_ag_dist(S,Q,[Ag|RAg],[d(Dist,Ag)|RDist])
   :  quad(S,Q,X1,Y1,X2,Y2) & init_pos(S,AgX,AgY)[source(Ag)]
   <- // get the distance between X1,Y1 and AgX,AgY
      jia.dist(X1,Y1,AgX,AgY,Dist);
      !calc_ag_dist(S,Q,RAg,RDist).
```

Implementation of Gold Searching

When the miner is free,[1] it has to systematically search for pieces of gold in its quadrant; it 'scans' its quadrant as illustrated in Figure 9.6. The plans for achieving this goal determine that the agent should start from the place where it last stopped searching for gold (represented by the belief last_checked(X,Y)), or if there is no information about previous location where the search stopped then the agent starts from the upper-left position within its quadrant (represented by the belief quadrant(X1,Y1,X2,Y2)[source(leader)]). The obstacles may prevent the agent from being in the exact location defined by the scan strategy, so the agent can skip to a close location, provided it is 'around' the target location. As the agent can perceive gold in all neighbouring cells, it can skip two lines when moving vertically.

```
/* beliefs and rules */

last_dir(null). // the last movement I did
free.

// next line is the bottom of the quadrant,
// if 2 lines below is too far
calc_new_y(AgY,QuadY2,QuadY2) :- AgY+2 > QuadY2.

// otherwise, the next line is 2 lines below
calc_new_y(AgY,_,Y) :- Y = AgY+2.
```

---

[1]The agent maintains a belief stating whether it is currently 'free' or not. Being free means that the agent is not committed to handling any piece of gold.

```
/* plans */

+free : last_checked(X,Y)      <- !around(X,Y).
+free : quadrant(X1,Y1,X2,Y2) <- !around(X1,Y1).
+free : true                   <- !wait_for_quad.

+!wait_for_quad : free & quadrant(_,_,_,_)
   <- -+free. // generate the event +free again
+!wait_for_quad : free
   <- .wait("+quadrant(X1,Y1,X2,Y2)", 500);
      !!wait_for_quad.

// if I am around the upper-left corner, move to upper-right corner
+around(X1,Y1) : quadrant(X1,Y1,X2,Y2) & free
   <- !prep_around(X2,Y1).

// if I am around the bottom-right corner, move to upper-left corner
+around(X2,Y2) : quadrant(X1,Y1,X2,Y2) & free
   <-!prep_around(X1,Y1).

// if I am around the right side, move to left side two lines below
+around(X2,Y) : quadrant(X1,Y1,X2,Y2) & free
   <- ?calc_new_y(Y,Y2,YF);
      !prep_around(X1,YF).

// if I am around the left side, move to right side two lines below
+around(X1,Y) : quadrant(X1,Y1,X2,Y2) & free
   <- ?calc_new_y(Y,Y2,YF);
      !prep_around(X2,YF).

// last "around" was none of the above, goes back to my quadrant
+around(X,Y) : quadrant(X1,Y1,X2,Y2) & free & Y <= Y2 & Y >= Y1
   <- !prep_around(X1,Y).

+around(X,Y) : quadrant(X1,Y1,X2,Y2) & free & X <= X2 & X >= X1
   <-!prep_around(X,Y1).

+!prep_around(X,Y) : free <- -around(_,_); -last_dir(_); !around(X,Y).

+!around(X,Y)
   :  // I am around to some location if I am near it or
      // the last action was "skip"
      // (meaning that there are no paths to there)
      (pos(AgX,AgY) & jia.neighbour(AgX,AgY,X,Y)) | last_dir(skip)
   <- +around(X,Y).
```

```
+!around(X,Y) : not around(X,Y)
  <- !next_step(X,Y);
     !!around(X,Y).
+!around(X,Y) : true
  <- !!around(X,Y).


+!next_step(X,Y)
  :  pos(AgX,AgY)
  <- jia.get_direction(AgX, AgY, X, Y, D);
     -+last_dir(D);
     D.
+!next_step(X,Y) : not pos(_,_) // I still do not know my position
  <- !next_step(X,Y).
-!next_step(X,Y) : true          // failure handling, start again
  <- -+last_dir(null);
     !next_step(X,Y).
```

Implementation of Gold Handling

When a miner sees a piece of gold, three relevant plans can be considered applicable depending on its current beliefs:

```
@pcell[atomic]      // atomic: so as not to handle another
                    // event while handle gold is initialised
+cell(X,Y,gold)
  :  not carrying_gold & free
  <- -free;
     +gold(X,Y);
     !init_handle(gold(X,Y)).


@pcell2[atomic]
+cell(X,Y,gold)
  :  // I desire to handle another gold which
     // is farther than the one just perceived
     not carrying_gold & not free &
     .desire(handle(gold(OldX,OldY))) &
     pos(AgX,AgY) &
     jia.dist(X,Y,AgX,AgY,DNewG) &
     jia.dist(OldX,OldY,AgX,AgY,DOldG) &
     DNewG < DOldG
  <- +gold(X,Y);
     .drop_desire(handle(gold(OldX,OldY)));
     .broadcast(untell, committed_to(gold(OldX,OldY)));
     .broadcast(tell,gold(OldX,OldY));
     !init_handle(gold(X,Y)).
```

```
+cell(X,Y,gold)
  :  not gold(X,Y) & not committed_to(gold(X,Y))
  <- +gold(X,Y);
     .print("Announcing ",gold(X,Y)," to others");
     .broadcast(tell,gold(X,Y)).
```

The first plan is applicable when the miner is not carrying gold and is free. The plan execution consists of removing the belief that it is free, adding a belief that there is gold at that location, and creating a goal to handle that particular piece of gold.

The second plan is applicable when the miner is also not carrying gold but is not free because it is going to position OldX, OldY to collect some gold there. In this case, it prefers the gold just found if it is closer, so the agent: (i) drops the previous intention; (ii) announces the availability of gold at the 'old' location to the other agents (this will trigger again the allocation protocol in Figure 9.8); and (iii) creates a goal to handle the piece of gold it has just found.

If none of the above plans is applicable (i.e. the agent is carrying gold), the third alternative plan is used to announce the gold location to others agents, starting the allocation protocol as before.

The code below implements the gold allocation protocol from the miner agent perspective (Figure 9.8). When the agent receives a message with some gold position, if it is free, it sends a bid based on its Manhattan distance to the gold; otherwise, it sends a very high bid. When some gold is allocated by the leader to an agent, that agent will intend to handle that piece of gold if the agent is still free. Note that this is not an optimal strategy: we have not as yet dealt with the possibility that reallocating tasks to the agents that are already committed (i.e. no longer free) might lead to a better overall task allocation.

```
// someone else sent me a gold location
+gold(X1,Y1)[source(A)]
   :  A \== self &
      not allocated(gold(X1,Y1),_) & // The gold was not allocated yet
      not carrying_gold &            // I am not carrying gold
      free &                         // and I am free
      pos(X2,Y2) & .my_name(Me)
   <- jia.dist(X1,Y1,X2,Y2,D);       // bid
      .send(leader,tell,bid(gold(X1,Y1),D,Me)).

// bid high as I'm not free
+gold(X1,Y1)[source(A)]
   :  A \== self & .my_name(Me)
   <- .send(leader,tell,bid(gold(X1,Y1),10000,Me)).

// gold allocated to me
@palloc1[atomic]
```

```
+allocated(Gold,Ag)[source(leader)]
  :  .my_name(Ag) & free // I am still free
  <- -free;
     !init_handle(Gold).

// some gold was allocated to me, but I cannot
// handle it anymore, re-announce
@palloc2[atomic]
+allocated(Gold,Ag)[source(leader)]
  :  .my_name(Ag) & not free // I am no longer free
  <- .broadcast(tell,Gold).

// someone else picked up the gold I am going to get,
// so drop the intention and chose another gold
@ppgd[atomic]
+picked(G)[source(A)]
  :  .desire(handle(G)) | .desire(init_handle(G))
  <- .abolish(G);
     .drop_desire(handle(G));
     !!choose_gold.

// someone else picked up a gold I know about,
// remove from my belief base
+picked(gold(X,Y))
  <- -gold(X,Y)[source(_)].
```

The gold allocation code in the leader agent is:

```
+bid(Gold,D,Ag)
  :  .count(bid(Gold,_,_),3)  // three bids were received
  <- !allocate_miner(Gold);
     .abolish(bid(Gold,_,_)).

+!allocate_miner(Gold)
  <- .findall(op(Dist,A),bid(Gold,Dist,A),LD);
     .min(LD,op(DistCloser,Closer));
     DistCloser < 10000;
     // Allocate the gold to the closest agent
     .broadcast(tell,allocated(Gold,Closer)).
```

Finally, the plans to achieve the goal of handling a found piece of gold are shown below. The plan initially drops the goal of finding gold (exploration behaviour), moves the agent to the gold position, picks up the gold, announces to others that the gold has been collected so they do not try to fetch this gold (to avoid agents moving to pieces of gold that are no longer there), retrieves the depot location from the belief base, moves the agent to the depot, drops the gold, and finally chooses another piece of gold to pursue. In case the handle-gold plan fails (e.g. because the gold disappeared due to the environment being dynamic), the event `-!handle(G)` is created and the second plan is selected. This plan just removes the information about that gold location from the belief base and chooses another piece of gold to be collected. The `choose_gold` plans find the nearest known gold and create a goal to handle it; if no gold location is known, the agent becomes free and resumes the gold-searching behaviour.

```
@pih1[atomic]
+!init_handle(Gold)
   :  .desire(around(_,_))
   <- .drop_desire(around(_,_));
      !init_handle(Gold).
@pih2[atomic]
+!init_handle(Gold)
   :  pos(X,Y)
   <- -+last_checked(X,Y);
      !!handle(Gold). // use !! so that "handle"
                      // isn't performed as atomic

+!handle(gold(X,Y))
   :  not free
   <- .broadcast(tell, committed_to(gold(X,Y)));
      !pos(X,Y);
      !ensure(pick,gold(X,Y));
      .broadcast(tell,picked(gold(X,Y)));
      ?depot(_,DX,DY);
      !pos(DX,DY);
      drop;
      -gold(X,Y)[source(_)];
      !!choose_gold.

// if ensure(pick/drop) failed, pursue another gold
-!handle(G)  : G
   <- -G[source(_)];
      !!choose_gold.
-!handle(_)
   <- !!choose_gold.

// no known gold to choose from...
// ... become free again to search for gold
+!choose_gold
   :  not gold(_,_)
   <- -+free.
```

```
// Finished one gold, but others left
// find the closest gold among the known options
// that nobody else committed to
+!choose_gold
  :  gold(_,_)
  <- .findall(gold(X,Y),gold(X,Y),LG);
     !calc_gold_distance(LG,LD);
     .length(LD,LLD); LLD > 0;
     .print("Uncommitted gold distances: ",LD,LLD);
     .min(LD,d(D,NewG));
     .print("Next gold is ",NewG);
     !!handle(NewG).
-!choose_gold
  <- -+free.

+!calc_gold_distance([],[]).
+!calc_gold_distance([gold(GX,GY)|R],[d(D,gold(GX,GY))|RD])
  :  pos(IX,IY) & not committed_to(gold(GX,GY))
  <- jia.dist(IX,IY,GX,GY,D);
     !calc_gold_distance(R,RD).
+!calc_gold_distance([_|R],RD)
  <- !calc_gold_distance(R,RD).

+!ensure(pick,G)
  :  pos(X,Y) & cell(X,Y,gold)
  <- pick;
     ?carrying_gold.
     // fails if there's no gold there or not carrying_gold after
     // pick!
     // handle(G) will "catch" this failure.

+!pos(X,Y) : pos(X,Y).
+!pos(X,Y) : not pos(X,Y)
  <- !next_step(X,Y);
     !pos(X,Y).
```

It is important to note that AgentSpeak is the language used to define the high-level (practical) reasoning of the agents. The AgentSpeak code for the team of gold miners, in our opinion, is a quite elegant solution, being declarative, goal-based (i.e. based on the BDI architecture), and also neatly allowing agents to have long-term goals while reacting to changes in the environment. Some of the advantages of using *Jason* are the support for high-level (speech-act based) communication, transparent integration with the contest server and support to the use of existing Java code through internal actions (e.g. for the A* algorithm). Although not a 'purely' declarative, logic-based approach, the combination of both declarative and legacy code allowed for an efficient solution without compromising the high-level declarative part of the implementation.

In future versions of this team, we plan to avoid the use of centralised negotiation (which has the leader as a single point of failure) and to use $\mathcal{M}oise^+$ [56] to create an organisation with the specification of the roles in the system (we say more about $\mathcal{M}oise^+$ in Section 11.2). In our original strategy, there was yet another role which was that of the 'courier'; where the depot happens to be in a position too far from the some of the quadrants, the courier will help carry to the depot pieces of gold from agents that are in the more distant quadrants; this is yet to be implemented. We also plan to experiment with DCOP algorithms for optimal allocation of agents to collect pieces of gold.

9.2 Case Study II: Electronic Bookstore

As mentioned in the introduction of this chapter, the electronic bookstore application is the example used throughout the book *Developing Intelligent Agent Systems* [75]. In this section, we do not describe the implementation of the whole system, since we are particularly interested in illustrating the integration of *Jason* with various technologies.

The main goal of this application is to sell books in an online system. To achieve this goal, among all identified functionalities available in the specification presented in [75], we selected the following to give as implementation examples:

Online interaction: manages the interaction with a single user, via the website.

Book finding: locates book information according to a given description.

Purchasing: manages online book sales.

Stock management: keeps track of the available stock.

These functionalities are distributed among three agents: sales assistant (online interaction, purchasing), delivery manager (purchasing) and stock manager (stock management, book finding). We also add an additional agent to simplify the implementation: the login agent. Based on a database of registered users, it validates the login of an user in the system and, if approved, creates the corresponding sales assistant for that user. Thus, each user logged on to the system has one sales assistant agent for her/himself.

All the above agents are developed in AgentSpeak and therefore can interact using *Jason*'s support for communication. However, the web application is not a *Jason* agent, and yet has to interact with the other agents. These agents may also be distributed in several different network hosts. Therefore, the centralised infrastructure cannot be used in this case; the Saci infrastructure was selected to

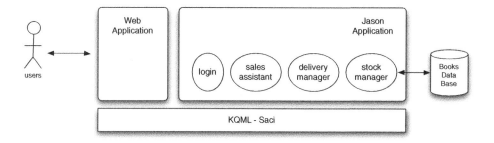

Figure 9.9 Overall architecture of the bookstore application.

provide distribution and (KQML-based) communication among heterogeneous agents. Figure 9.9 shows the overall architecture of the system. The next two sections describe each part – the web application and *Jason* agents – of this architecture.

Web Application Agent

The whole web application was developed with Java Server Pages (JSP) and thus runs on an application server. An example of a search result in this web site is shown in the screenshot in Figure 9.10. To allow this application to send and receive KQML-like messages, the application has a Java bean component (a Saci mailbox) created when the application starts and shared by all pages.

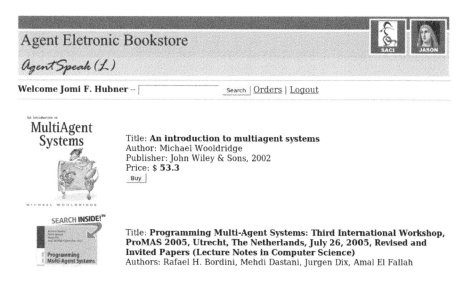

Figure 9.10 Electronic bookstore example web page.

For example, the `login.jsp` page receives the user name and password filled in by the user in a form, sends a message to the login agent to validate the user, and if it is valid forwards the user to its personal page. Part of the code for this page is:

```
...
// Declaration of the Saci mail box bean.
<jsp:useBean id="mbox" scope="application"
                     class="saci.MBoxSAg" />
...

// get parameters from the from
String userName = request.getParameter("txtUser");
String password = request.getParameter("txtPassword");

// creates a KQML message to send to the 'login' agent
Message m =
    new Message("(askOne :receiver login :language AgentSpeak)");
m.put("content", "user_logon("+userName+",\""+password+"\")");

// sends a message and waits for the answer
Message r = mbox.ask(m);

String content = r.get("content").toString();
if (content.startsWith("client")) {
   Structure s = Structure.parse(content);
   String name = s.getTerm(0).toString();
   session.setAttribute("username",userName);
   session.setAttribute("name",name);
   %>
   <jsp:forward page="userPage.jsp" />
   <%
}
...
```

To illustrate the communication between the web application (more precisely, its `login.jsp` page) and the login agent, let us suppose that the user logging in to the systems is 'bob' with password 'test'; the message sent to the login agent will be:

```
(askOne :sender    web
        :receiver  login
        :reply-with id1
        :language  AgentSpeak
        :content   "user_logon(bob,\"test\")")
```

The sender and reply-with slots are automatically added by the **ask** method of the mailbox. This method also waits for the answer which, if the password is correct, is:

```
(tell    :sender        login
         :receiver      web
         :in-reply-to   id1
         :language      AgentSpeak
         :content       "client(\"Bob Spengler\")")
```

The web page finally registers the user name and full name in the user's session so that other pages can access this information, and forwards it to the user's personal page. The remaining pages of the application do not send messages to the login agent but to the sales assistant created for the user (there is a guest sales assistant for users not logged in), as in the following excerpt of the search page:

```
<jsp:useBean id="mbox" scope="application" class="saci.MBoxSAg" />

<%
String key = request.getParameter("key");

if  (key != null && key.length() > 0) {
    String userName = session.getAttribute("username").toString();

    Message m  = new Message("(askOne)");
    m.put("receiver", userName);
    m.put("content", "find_book(\""+key+"\",L)");
    Message answer = mbox.ask(m);

    // the answer is a list of books code
    Literal  content = Literal.parseLiteral(
                                answer.get("content").toString());
    ListTerm lbooks = (ListTerm)content.getTerm(1);
...
```

In the search protocol, the message sent by the web application is

```
(askOne :sender        web
        :receiver       bob
        :reply-with     id2
        :language       AgentSpeak
        :content        "find_book(\"Agent\",L)")
```

and the answer is something like

```
(tell    :sender        bob
         :receiver       web
```

```
:in-reply-to id2
:language    AgentSpeak
:content     "find_book(\"Agent\",[3,4,30])")
```

where [3,4,30] is a list of book references for which the keyword 'Agent' was found.

Jason Agents

While the web application focuses on the interface, the functionalities of the system are implemented as agents in *Jason*. The system starts with four agents: the three agents described earlier (and defined in [75]) and a default sales assistant, called guest, for guest users (i.e. those who are not logged on to the system). More sales assistant agents are dynamically created when users log in. These agents use Saci as infrastructure and so can be distributed in different hosts as defined in the following system configuration:

```
MAS electronicBookstore {
   infrastructure:
      Saci

   agents:
      login
         beliefBaseClass jason.bb.TextPersistentBB
         at "web.server.com"; // the host where this agent starts

      guest salesAssistant at "web.server.com";

      deliveryManager
         beliefBaseClass jason.bb.TextPersistentBB
         at "int.server.com";

      stockManager
         beliefBaseClass jason.bb.JDBCPersistentBB(
            "org.hsqldb.jdbcDriver", // driver for HSQLDB
            "jdbc:hsqldb:bookstore", // URL connection
            "sa", // user
            "",   // password
            "[book(7),publisher(2),book_stock(2),
              author(2),book_author(2),
              topic(2),book_topic(2)]")
         at "bd.server.com";
}
```

In this configuration, text files are used to persist the belief base of the login and delivery manager agents. The stock manager uses a database to persist and

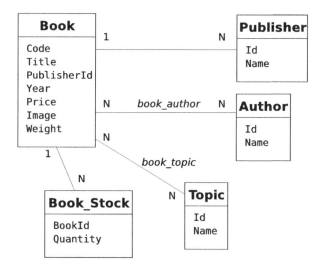

Figure 9.11 Entity-relationship model of the stock manager database.

access the information about book title, publisher, author, topic and the relations between them (the features described in Section 7.4 are used to bind the beliefs to tables of a relational database). The entity-relationship model of this database is specified in Figure 9.11. We will not include here all the AgentSpeak source code of the agents, since most of them are quite simple and are available at the web page for this book (http://jason.sf.net/jBook). Below, we comment on parts of the code for the stock manager and login agents.

The stock manager has access to the book database (modelled as in Figure 9.11) and thus is responsible for finding a book given a keyword, for instance. When it receives a message, with content find_book(Key,BookList), from the web application asking for a search, the default implementation of KQML tries to find a belief that unifies with the content (see Chapter 6 for more information on communication). Of course, there is no such belief. In this case, an event +?find_book(Key,BookList) is generated and handled by the last plan in the code below. Note that this last plan makes extensive use of rules and backtracking to implement the 'find book' functionality of this agent.

```
// find books by some author
//   the first argument is the author to be found and
//   the second is a book's code
find_by_author(SearchAuthor,Code)
     :- book(Code, _, _, _, _, _, _) & // for all books in DB
        book_author(Code,AuthorId) &   // for all its authors
        author(AuthorId,Name) &        // get author's name
        .substring(SearchAuthor,Name).
```

```
// find book by its title
find_by_title(SearchTitle,Code)
    :-  book(Code, Title, _, _, _, _, _) &
        .substring(SearchTitle,Title).

// the plan for askOne find_book
+?find_book(Key, L)
   <- .findall(CA,find_by_author(Key,CA),LA);
      .findall(CT,find_by_title(Key,CT),LT);
      .concat(LA,LT,L).
```

The login agent implements the login protocol described in the introduction of this case study. It has to create a new agent for each user that logs in to the system, and has to kill this agent when the user logs out.

```
+!kqml_received(S, askOne, user_logon(UserName, Password), M)
   :  // S is the sender and M is the reply-with id
      client(UserName, Password, Name, _, _, _, _, _, _, _)
   <- // creates an agent for this user
      .create_agent(
          UserName,                        // agent name
          "salesAssistant.asl",     // source code
          [beliefBaseClass(jason.bb.TextPersistentBB)]); // options
      .send(S, tell, client(Name), M).

+!kqml_received(S, askOne, user_logon(UserName, Password), M)
   <-  .send(S, tell, error("Username or password invalid!"), M).

+user_logout(UserName)[source(web)]
   <- .kill_agent(UserName).
```

The first two plans exemplify customisations of the *Jason* default communication semantics. The default semantics for the askOne is not suitable for the desired login protocol, where the answer (client(Name)) does not unify the ask content (user_logon(UserName,Password)). Of course, the login protocol could be changed to fit the *Jason* usual features. However, the intention here is to demonstrate how a KQML-compliant agent that was not implemented in *Jason* can interact with a *Jason* agent. With the customisation, when an askOne user_logon message arrives, these two plans are relevant. The first is applied when the user and password are correct, according to a belief client(...) that is used to represent all valid users; the second plan is used otherwise. To dynamically create a new sales assistant agent, the internal action .create_agent is used (see page 252 for more detail on this internal action and its arguments). Similarly, the internal action .kill_agent is used in the last plan to kill the sales assistant agent when the user logs out.

10

Formal Semantics

While the emphasis in previous chapters has been on how to program in AgentSpeak using *Jason*, this chapter is concerned more with the theoretical foundations of the language. (As such, it is not necessary reading for readers who are only interested in programming; it is more important for those who want to do research related to AgentSpeak.) We present the semantics for the core AgentSpeak language. An earlier version of the operational semantics for AgentSpeak appeared in [10]. The semantics rules for communication appeared in [70] and were later improved and extended in [100], which is the version used as the basis for this chapter.

It should be fairly clear that this is a much simplified version of the language interpreted by *Jason* (cf. the grammar in Appendix A.1), but it contains all the fundamental aspects of the language, and is therefore more appropriate for giving formal semantics. The formal semantics will be given for the language defined by the following grammar:

$$
\begin{array}{lll}
ag & ::= & bs \quad ps \\
bs & ::= & b_1 \ldots b_n & (n \geq 0) \\
ps & ::= & p_1 \ldots p_n & (n \geq 1) \\
p & ::= & te : ct \leftarrow h \\
te & ::= & +at \quad | \quad -at \quad | \quad +g \quad | \quad -g \\
ct & ::= & ct_1 \quad | \quad \top \\
ct_1 & ::= & at \quad | \quad \neg at \quad | \quad ct_1 \wedge ct_1 \\
h & ::= & h_1; \top \quad | \quad \top \\
h_1 & ::= & a \quad | \quad g \quad | \quad u \quad | \quad h_1; h_1 \\
at & ::= & \mathrm{P}(t_1, \ldots, t_n) & (n \geq 0) \\
& | & \mathrm{P}(t_1, \ldots, t_n)[s_1, \ldots, s_m] & (n \geq 0, m > 0)
\end{array}
$$

$$
\begin{array}{llll}
s & ::= & \text{percept} \mid \text{self} \mid & id \\
a & ::= & \text{A}(t_1, \ldots, t_n) & (n \geq 0) \\
g & ::= & !at \quad \mid \quad ?at & \\
u & ::= & +b \quad \mid \quad -at &
\end{array}
$$

In this grammar, we use *ag* to denote an agent, which is formed by a set of beliefs *bs* and a set of plans *ps*. We use *b* to denote an individual belief, which is a ground (first-order) atomic formula, and we use *at* for atomic formulæ which might not be ground. We use *p* for an individual plan, *te* for a triggering event, *ct* for a plan context, *h* for a plan body (and ⊤ is the empty plan body), *s* for the information source (*id* ranges over labels representing the agents in the system), *a* for actions, *g* for goals and *u* for belief updates which are essentially changes in the agent's 'mental notes'. Note that this simplified language assumes that only annotations about sources of information exist (because they are the only ones with special meaning for the language interpretation), so they are not used within a source(s) predicate as in *Jason*.

We use operational semantics [76] to define the semantics of AgentSpeak; this approach is widely used for giving semantics to programming languages and studying their properties. The operational semantics is given by a set of rules that define a transition relation between configurations $\langle ag, C, M, T, s \rangle$ where:

- An agent program *ag* is formed by a set of beliefs *bs* and a set of plans *ps* (as defined in the grammar above).

- An agent's circumstance *C* is a tuple $\langle I, E, A \rangle$ where:

    - *I* is a set of *intentions* $\{i, i', \ldots\}$; each intention *i* is a stack of partially instantiated plans.
    - *E* is a set of *events* $\{(te, i), (te', i'), \ldots\}$. Each event is a pair (te, i), where *te* is a triggering event and *i* is an intention (a stack of plans in the case of an internal event, or the empty intention ⊤ in case of an external event). When the belief-update function (which is not part of the AgentSpeak interpreter but rather of the agent's overall architecture) updates the belief base, the associated events – i.e. additions and deletions of beliefs – are included in this set. These are called *external* events; internal events are generated by additions or deletions of goals.
    - *A* is a set of *actions* to be performed in the environment. An action expression included in this set tells other architectural components to actually perform the respective action on the environment, thus changing it.

- *M* is a tuple $\langle In, Out, SI \rangle$ whose components register the following aspects of communicating agents:

- *In* is the mail inbox: the system includes all messages addressed to this agent in this set. Elements of this set have the form $\langle mid, id, ilf, cnt \rangle$, where *mid* is a message identifier, *id* identifies the sender of the message, *ilf* is the illocutionary force of the message and *cnt* its content, which can be an (or a set of) AgentSpeak predicate(s) or plan(s), depending on the illocutionary force of the message.

- *Out* is where the agent posts all messages it wishes to send to other agents; the underlying multi-agent system mechanism makes sure that messages included in this set are sent to the agent addressed in the message. Messages here have exactly the same format as above, except that here *id* refers to the agent to which the message is to be sent.

- *SI* is used to keep track of intentions that were suspended due to the processing of communication messages; this is explained in more detail in the next section, but the intuition is as follows: intentions associated with illocutionary forces that require a reply from the interlocutor are suspended, and they are only resumed when such reply has been received.

- It helps to use a structure which keeps track of temporary information that is required in subsequent stages within a single reasoning cycle. T is the tuple $\langle R, Ap, \iota, \varepsilon, \rho \rangle$ with such temporary information; it has as components:

  - R for the set of *relevant plans* (for the event being handled);
  - Ap for the set of *applicable plans* (the relevant plans whose contexts are believed true);
  - ι, ε and ρ record a particular intention, event and applicable plan (respectively) being considered along the execution of one reasoning cycle.

- The current step s within an agent's reasoning cycle is symbolically annotated by $s \in \{\mathsf{ProcMsg, SelEv, RelPl, ApplPl, SelAppl, AddIM, SelInt, ExecInt, ClrInt}\}$, standing for: processing a message from the agent's mail inbox, selecting an event from the set of events, retrieving all relevant plans, checking which of those are applicable, selecting one particular applicable plan (the intended means), adding the new intended means to the set of intentions, selecting an intention, executing the selected intention and clearing an intention or intended means that may have finished in the previous step.

In the interests of readability, we adopt the following notational conventions in our semantic rules:

- If C is an AgentSpeak agent circumstance, we write C_E to make reference to the component E of C. Similarly for all the other components of a configuration.

- We write $T_\iota = \_$ (the underscore symbol) to indicate that there is no intention presently being considered in that reasoning cycle. Similarly for T_ε and T_ρ.

- We write $i[p]$ to denote the intention that has plan p on top of intention i.

The AgentSpeak interpreter makes use of three *selection functions* that are defined by the agent programmer. The selection function $\mathcal{S}_\varepsilon$ selects an event from the set of events C_E; the selection function $\mathcal{S}_O$ selects an applicable plan given a set of applicable plans; and $\mathcal{S}_\mathcal{I}$ selects an intention from the set of intentions C_I (these functions were explained in Chapter 4.1 and recall that *Jason* allows these methods to be overridden, as shown in Chapter 7.2). Formally, all the selection functions an agent uses are also part of its configuration (as is the social acceptance function that we mention below). However, as they are defined by the agent programmer at design time and do not (in principle) change at run time, we avoid including them in the configuration, for the sake of readability.

Further, we define some auxiliary syntactic functions to help the presentation of the semantics. If p is a plan of the form $te : ct \leftarrow h$, we define $\mathsf{TrEv}(p) = te$ and $\mathsf{Ctxt}(p) = ct$. That is, these projection functions return the triggering event and the context of the plan, respectively. The TrEv function can also be applied to the head of a plan rather than the whole plan, but works similarly in that case.

Next, we need to define the specific (limited) notion of logical consequence used here. We assume the existence of a procedure that computes the most general unifier of two literals (as usual in logic programming), and with this define the logical consequence relation $\models$ that is used in the definitions of the auxiliary functions for checking for relevant and applicable plans, as well as executing test goals. Given that we have extended the syntax of atomic formulæ so as to include annotations of the sources for the information symbolically represented by it, we also need to define $\models$ in our particular framework, as follows.

Definition 10.1 *We say that an atomic formula at_1 with annotations $s_{11}, \ldots , s_{1n}$ is a logical consequence of a set of ground atomic formulæ bs, written $bs \models at_1[s_{11}, \ldots , s_{1n}]$ if, and only if, there exists $at_2[s_{21}, \ldots , s_{2m}] \in bs$ such that (i) $at_1\theta = at_2$, for some most general unifier θ, and (ii) $\{s_{11}, \ldots , s_{1n}\} \subseteq \{s_{21}, \ldots , s_{2m}\}$.*

The intuition is that not only should predicate at unify with some of the predicates in bs (i), but also that all specified sources of information for at should be corroborated in bs (ii). Thus, for example, $\mathtt{p(X)[ag_1]}$ follows from $\{\mathtt{p(t)[ag_1,ag_2]}\}$, but $\mathtt{p(X)[ag_1,ag_2]}$ does *not* follow from $\{\mathtt{p(t)[ag_1]}\}$. More concretely, if a plan requires, to be applicable, that a drowning person was explicitly perceived rather than communicated by another agent (which can be represented by `drowning(Person)[percept]`), that follows from a belief `drowning(man)[percept,passerby]` (i.e. that this was both perceived and communicated by a passerby). On the other hand, if the required context was that two independent sources provided the information, say `cheating(Person)[witness1,witness2]`, this cannot be inferred from a belief `cheating(husband)[witness1]`.

In order to make some semantic rules more readable, we use two operations on belief bases (i.e. sets of annotated predicates). We use $bs' = bs + b$ to say that bs' is as bs except that $bs' \models b$. Similarly $bs' = bs - b$ means that bs' is as bs except that $bs' \not\models b$.

A plan is considered *relevant* in relation to a triggering event if it has been written to deal with that event. In practice, this is checked by trying to unify the triggering event part of the plan with the triggering event within the event that has been selected for handling in that reasoning cycle. In the definition below, we use the logical consequence relation defined above to check if a plan's triggering event unifies with the event that has occurred. To do this, we need to extend the $\models$ relation so that it also applies to triggering events instead of predicates. In fact, for the purposes here, we can consider that any operators in a triggering event (such as '+' or '!') are part of the predicate symbol or, more precisely, let at_1 be the predicate (with annotation) within triggering event te_1 and at_2 the one within te_2, then $\{te_2\} \models te_1$ if, and only if, $\{at_2\} \models at_1$ and, of course, the operators prefixing te_1 and te_2 are exactly the same. Because of the requirement of inclusion of annotations, the converse may not be true.

Definition 10.2 *Given the plans ps of an agent and a triggering event te, the set* RelPlans(ps, te) *of relevant plans is given as follows:*

$$RelPlans(ps, te) = \{(p, \theta) \mid p \in ps \text{ and } \theta \text{ is s.t. } \{te\} \models TrEv(p)\theta\}.$$

The intuition here, regarding annotations, is as follows. The programmer should include in the annotations of a plan's triggering event all the sources that must have generated the event for that plan to be relevant (or include no annotation if the source of information is not important for the plan to be relevant). For the plan to be relevant, it therefore suffices for the annotations in the plan's triggering event to be a subset of those in the event that occurred. A plan with triggering event $+!p(X)[s]$ is relevant for an event $\langle +!p(t)[s, t], \top \rangle$ since RelPlans requires that $\{p(t)[s, t]\} \models p(X)[s]\theta$ (for some most general unifier θ), which in turn requires that $\{s\} \subseteq \{s, t\}$. As a consequence, for a plan with a triggering event that has no annotations (e.g., $+!p(X)$) to be relevant for a particular event (say, $\langle +!p(t)[ag_1], i \rangle$) requires only that the predicates unify in the usual sense since $\{\} \subseteq S$, for any set S.

A plan is *applicable* if it is relevant and its context is a logical consequence of the agent's beliefs. Again we need to extend slightly the definition of $\models$ given above. A plan's context is a conjunction of literals (l is either at or $\neg at$). We can say that $bs \models l_1 \wedge \ldots \wedge l_n$ if, and only if, $bs \models l_i$ if l_i is of the form at, and $bs \not\models l_i$ if l_i is of the form $\neg at$, for $1 \leq i \leq n$. The auxiliary function for applicable plans is formalised as follows.

Definition 10.3 *Given a set of relevant plans R and the beliefs bs of an agent, the set of applicable plans* AppPlans(bs, R) *is defined as follows:*

$$\textsf{AppPlans}(bs, R) = \{(p, \theta' \circ \theta) \mid (p, \theta) \in R \text{ and } \theta' \text{ is s.t. } bs \models \textsf{Ctxt}(p)\theta\theta'\}.$$

We also need an auxiliary function to help in the semantic rule that is used when the agent is executing a test goal. The evaluation of a test goal $?at$ consists in testing if the formula at is a logical consequence of the agent's beliefs. The auxiliary function returns a set of most general unifiers all of which make the formula at a logical consequence of a set of formulæ bs, as follows.

Definition 10.4 *Given a set of formulæ bs and a formula at, the set of substitutions* Test(bs, at) *produced by testing at against bs is defined as follows:*

$$\textsf{Test}(bs, at) = \{\theta \mid bs \models at\theta\}.$$

Next, we present the rules which define the operational semantics of the reasoning cycle of AgentSpeak. Before we do so, it helps in understanding the semantics if we consider all the stages of a AgentSpeak reasoning cycle. The graph in Figure 10.1 shows all possible transitions between the various steps in an agent's reasoning cycle, helping to give an overview of an agent reasoning cycle as defined by the semantics. The step labels stand for: processing received messages (**ProcMsg**), selecting an event from the set of events (**SelEv,**) retrieving all relevant plans (**RelPl**), checking which of those are applicable (**ApplPl**), selecting one particular applicable plan – the intended means (**SelAppl**), adding the new intended means to the set of intentions (**AddIM**), selecting an intention (**SelInt**),

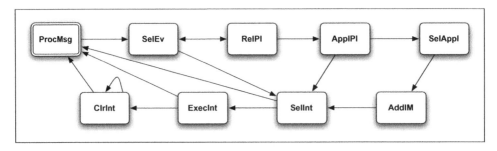

Figure 10.1 Possible state transitions within one reasoning cycle.

executing the selected intention (**ExecInt**), and clearing an intention or intended means that may have finished in the previous step (**ClrInt**). Note that the **ProcMsg** stage is the one used for giving semantics of communication in AgentSpeak; the semantic rules for communication are given separately in Section 10.3.

10.1 Semantic Rules

In the general case, an agent's initial configuration is $\langle ag, C, M, T, \mathsf{ProcMsg}\rangle$, where ag is as given by the agent program, and all components of C, M and T are empty. Note that a reasoning cycle starts with processing received messages (**ProcMsg**), according to a recent extension of the semantics [100], and included below in Section 10.3. After that, an event selection (**SelEv**) is made, starting the reasoning cycle as originally defined for the language, which is the part of the semantics presented below.

Event selection: the rule below assumes the existence of a selection function $\mathcal{S}_{\mathcal{E}}$ that selects events from a set of events E. The selected event is removed from E and it is assigned to the ε component of the temporary information. Rule **SelEv$_2$** skips to the intention execution part of the cycle, if there is no event to handle.

$$\frac{\mathcal{S}_{\mathcal{E}}(C_E) = \langle te, i\rangle}{\langle ag, C, M, T, \mathsf{SelEv}\rangle \longrightarrow \langle ag, C', M, T', \mathsf{RelPl}\rangle} \qquad (\mathbf{SelEv_1})$$

$$\text{where:} \quad \begin{aligned} C'_E &= C_E \setminus \{\langle te, i\rangle\} \\ T'_\varepsilon &= \langle te, i\rangle \end{aligned}$$

$$\frac{C_E = \{\}}{\langle ag, C, M, T, \mathsf{SelEv}\rangle \longrightarrow \langle ag, C, M, T, \mathsf{SelInt}\rangle} \qquad (\mathbf{SelEv_2})$$

Relevant plans: rule **Rel$_1$** assigns the set of relevant plans to component T_R. Rule **Rel$_2$** deals with the situation where there are no relevant plans for an event; in that case, the event is simply discarded. In fact, an intention associated with that event might be completely discarded too; if there are no relevant plans to handle an event generated by that intention, it cannot be further executed. In practice this leads to activation of the plan failure mechanism, which we do not formalise here.

$$\frac{T_\varepsilon = \langle te, i\rangle \qquad \mathsf{RelPlans}(ag_{ps}, te) \neq \{\}}{\langle ag, C, M, T, \mathsf{RelPl}\rangle \longrightarrow \langle ag, C, M, T', \mathsf{ApplPl}\rangle} \qquad (\mathbf{Rel_1})$$

$$\text{where:} \quad T'_R = \mathsf{RelPlans}(ag_{ps}, te)$$

$$\frac{\mathsf{RelPlans}(ag_{ps}, te) = \{\}}{\langle ag, C, M, T, \mathsf{RelPl}\rangle \longrightarrow \langle ag, C, M, T, \mathsf{SelEv}\rangle} \quad (\mathbf{Rel_2})$$

An alternative approach for what to do if there are no relevant plans for an event is described in [6]. It assumes that in some cases, explicitly specified by the programmer, the agent will want to ask other agents what is the know-how they use to handle such events. The mechanism for plan exchange between AgentSpeak agents presented in [6] allows the programmer to specify which triggering events should generate attempts to retrieve external plans, which plans an agent agrees to share with others, what to do once the plan has been used for handling that particular event instance, and so on.

Applicable plans: the rule **Appl₁** assigns the set of applicable plans to the T_{Ap} component; rule **Appl₂** applies when there are no applicable plans for an event, in which case the event is simply discarded and that reasoning cycle carries on to check if there is any intention to be further executed. Again, in practice, this normally leads to the plan failure mechanism being used, rather then simply discarding the event (and possibly a whole intention with it).

$$\frac{\mathsf{AppPlans}(ag_{bs}, T_R) \neq \{\}}{\langle ag, C, M, T, \mathsf{ApplPl}\rangle \longrightarrow \langle ag, C, M, T', \mathsf{SelAppl}\rangle} \quad (\mathbf{Appl_1})$$

$$\text{where:} \quad T'_{Ap} \;=\; \mathsf{AppPlans}(ag_{bs}, T_R)$$

$$\frac{\mathsf{AppPlans}(ag_{bs}, T_R) = \{\}}{\langle ag, C, M, T, \mathsf{ApplPl}\rangle \longrightarrow \langle ag, C, M, T, \mathsf{SelInt}\rangle} \quad (\mathbf{Appl_2})$$

Selection of an applicable plan: this rule assumes the existence of a selection function $\mathcal{S}_{\mathcal{O}}$ that selects a plan from a set of applicable plans T_{Ap}. The selected plan is then assigned to the T_{ρ} component of the configuration.

$$\frac{\mathcal{S}_{\mathcal{O}}(T_{Ap}) = (p, \theta)}{\langle ag, C, M, T, \mathsf{SelAppl}\rangle \longrightarrow \langle ag, C, M, T', \mathsf{AddIM}\rangle} \quad (\mathbf{SelAppl})$$

$$\text{where:} \quad T'_{\rho} \;=\; (p, \theta)$$

Adding an intended means to the set of intentions: events can be classified as external or internal (depending on whether they were generated from the agent's perception, or whether they were generated by the previous execution of other plans, respectively). Rule **ExtEv** says that if the event ε is external (which is indicated by $\top$ in the intention associated with ε), a new intention is created and the only intended means in it is the plan p assigned to the ρ component. If the event is internal, rule **IntEv** says that the plan in ρ should be put on top of the intention associated with the event.

$$\frac{T_\varepsilon = \langle te, \top \rangle \qquad T_\rho = (p, \theta)}{\langle ag, C, M, T, \mathsf{AddIM} \rangle \longrightarrow \langle ag, C', M, T, \mathsf{SelInt} \rangle} \qquad \text{(ExtEv)}$$

$$\text{where:} \quad C'_I \;=\; C_I \cup \{\,[p\theta]\,\}$$

$$\frac{T_\varepsilon = \langle te, i \rangle \qquad T_\rho = (p, \theta)}{\langle ag, C, M, T, \mathsf{AddIM} \rangle \longrightarrow \langle ag, C', M, T, \mathsf{SelInt} \rangle} \qquad \text{(IntEv)}$$

$$\text{where:} \quad C'_I \;=\; C_I \cup \{\, i[(p\theta)]\,\}$$

Note that, in rule **IntEv**, the whole intention i that generated the internal event needs to be inserted back in C_I, with p as its top. This issue is related to suspended intentions and discussed further when we later present rule **AchvGl**.

Intention selection: rule **SelInt₁** assumes the existence of a function $\mathcal{S}_\mathcal{I}$ that selects an intention (i.e. a stack of plans) for processing next, while rule **SelInt₂** takes care of the situation where the set of intentions is empty (in which case, the reasoning cycle starts from the beginning).

$$\frac{C_I \neq \{\} \qquad \mathcal{S}_\mathcal{I}(C_I) = i}{\langle ag, C, M, T, \mathsf{SelInt} \rangle \longrightarrow \langle ag, C, M, T', \mathsf{ExecInt} \rangle} \qquad \text{(SelInt}_1\text{)}$$

$$\text{where:} \quad T'_\iota \;=\; i$$

$$\frac{C_I = \{\}}{\langle ag, C, M, T, \mathsf{SelInt} \rangle \longrightarrow \langle ag, C, M, T, \mathsf{ProcMsg} \rangle} \qquad \text{(SelInt}_2\text{)}$$

Executing an intended means: this group of rules express the effects of executing the body of a plan; each rule deals with one type of formula that can appear in a plan body. The plan to be executed is always the one on top of the intention that has been selected in the previous step; the specific formula to be executed is the one at the beginning of the body of that plan.

Actions: the action a in the body of the plan is added to the set of actions A. The action is removed from the body of the plan and the intention is updated to reflect this.

$$\frac{T_\iota = i[head \leftarrow a; h]}{\langle ag, C, M, T, \mathsf{ExecInt} \rangle \longrightarrow \langle ag, C', M, T', \mathsf{ClrInt} \rangle} \qquad \textbf{(Action)}$$

$$\text{where:} \quad \begin{aligned} C'_A &= C_A \cup \{a\} \\ T'_\iota &= i[head \leftarrow h] \\ C'_I &= (C_I \setminus \{T_\iota\}) \cup \{T'_\iota\} \end{aligned}$$

Achievement goals: this rule registers a new internal event in the set of events E. This event can then be eventually selected (see rule **SelEv**). When the formula being executed is a goal, the formula is not removed from the body of the plan, as in the other cases. This only happens when the plan used for achieving that goal finishes successfully; see rule **ClrInt$_2$**. The reasons for this are related to further instantiation of the plan as well as handling plan failure.

$$\frac{T_\iota = i[head \leftarrow !at; h]}{\langle ag, C, M, T, \mathsf{ExecInt} \rangle \longrightarrow \langle ag, C', M, T, \mathsf{ProcMsg} \rangle} \qquad \textbf{(AchvGl)}$$

$$\text{where:} \quad \begin{aligned} C'_E &= C_E \cup \{\langle +!at, T_\iota \rangle\} \\ C'_I &= C_I \setminus \{T_\iota\} \end{aligned}$$

Note how the intention that generated the internal event is removed from the set of intentions C_I; this captures the idea of *suspended intentions*. When the event with the achievement-goal addition is handled and a plan for it is selected (rule **IntEv**), the intention can be resumed by executing the plan for achieving that goal. If we have, in a plan body, '$!g; f$' (where f is any formula that can appear in plan bodies), this means that, before f can be executed, the state of affairs represented by goal g needs to be achieved (through the execution of some relevant, applicable plan). This newly added goal is treated as any other event, which means it will be placed in the set of events until it is eventually selected in a later reasoning cycle. Meanwhile, that plan (with formula f to be executed next) can no longer be executed, hence the whole intention (recall that an intention is stack of plans) is suspended by being placed, within an event, in the set of events and removed from the set of intentions. When a plan for achieving g has been selected, it is pushed on top of the suspended intention, which is then resumed (i.e. moved back to the set of intentions), according to rule **IntEv**. The execution of that intention proceeds with the plan at the top (in this case, for achieving g), and only when that plan is finished will f be executed (as it will be at the top of the intention again).

Test goals: these rules are used when a test goal formula $?at$ is to be executed. Rule **TestGl$_1$** is used when there is a set of substitutions that can make at a logical consequence of the agent's beliefs.[1] If the test goal succeeds, the substitution is

---

applied to the whole intended means, and the reasoning cycle can be continued. If that is not the case, it may be that the test goal is used as a triggering event of a plan, which is used by programmers to formulate more sophisticated queries. Rule **TestGl$_2$** is used in such case: it generates an internal event, which may trigger the execution of a plan, as for achievement goals. If to carry out a plan an agent is required to obtain information (at the time of actual execution of the plan) which is not directly available in its belief base, a plan for a test goal can be written which, for example, sends messages to other agents, or processes available data, so that the particular test goal can be concluded (producing an appropriate instantiation of logical variables). If an internal event is generated for the test goal being executed, the process is very similar to achievement goals, where the intention is suspended until a plan is selected to achieve the goal, as explained above.

$$\frac{T_\iota = i[head \leftarrow ?at; h] \qquad \mathsf{Test}(ag_{bs}, at) \neq \{\}}{\langle ag, C, M, T, \mathsf{ExecInt}\rangle \longrightarrow \langle ag, C', M, T, \mathsf{ClrInt}\rangle} \qquad \text{(TestGl}_1\text{)}$$

$$\begin{aligned}
\text{where:} \quad T'_\iota &= i[(head \leftarrow h)\theta] \\
&\quad \theta \in \mathsf{Test}(ag_{bs}, at) \\
C'_I &= (C_I \setminus \{T_\iota\}) \cup \{T'_\iota\}
\end{aligned}$$

$$\frac{T_\iota = i[head \leftarrow ?at; h] \qquad \mathsf{Test}(ag_{bs}, at) = \{\}}{\langle ag, C, M, T, \mathsf{ExecInt}\rangle \longrightarrow \langle ag, C', M, T, \mathsf{ClrInt}\rangle} \qquad \text{(TestGl}_2\text{)}$$

$$\begin{aligned}
\text{where:} \quad C'_E &= C_E \cup \{\langle +?at, T_\iota\rangle\} \\
C'_I &= C_I \setminus \{T_\iota\}
\end{aligned}$$

Updating beliefs: rule **AddBel** simply adds a new event to the set of events E. The formula $+b$ is removed from the body of the plan and the set of intentions is updated properly. Rule **DelBel** works similarly. In both rules, the set of beliefs of the agent should be modified in such a way that either the predicate b (with annotation `self`) is included in the new set of beliefs (rule **AddBel**) or it is removed from there (rule **DelBel**). Note that a request to delete beliefs can have variables (at), whilst only ground atoms (b) can be added to the belief base.

$$\frac{T_\iota = i[head \leftarrow +b; h]}{\langle ag, C, M, T, \mathsf{ExecInt}\rangle \longrightarrow \langle ag', C', M, T', \mathsf{ClrInt}\rangle} \qquad \text{(AddBel)}$$

$$\begin{aligned}
\text{where:} \quad ag'_{bs} &= ag_{bs} + b[\texttt{self}] \\
C'_E &= C_E \cup \{\langle +b[\texttt{self}], \top\rangle\} \\
T'_\iota &= i[head \leftarrow h] \\
C'_I &= (C_I \setminus \{T_\iota\}) \cup \{T'_\iota\}
\end{aligned}$$

$$\frac{T_\iota = i[head \leftarrow -at; h]}{\langle ag, C, M, T, \mathsf{ExecInt} \rangle \longrightarrow \langle ag', C', M, T, \mathsf{ClrInt} \rangle} \quad \text{(DelBel)}$$

$$\begin{aligned}
\text{where:} \quad ag'_{bs} &= ag_{bs} - at[\texttt{self}] \\
C'_E &= C_E \cup \{\langle -at[\texttt{self}], \top \rangle\} \\
T'_\iota &= i[head \leftarrow h] \\
C'_I &= (C_I \setminus \{T_\iota\}) \cup \{T'_\iota\}
\end{aligned}$$

Clearing intentions: finally, the following rules remove empty intended means or intentions from the set of intentions. Rule **ClrInt$_1$** simply removes a whole intention when there is nothing else to be executed in that intention. Rule **ClrInt$_2$** clears the remainder of the plan with an empty body currently at the top of a (non-empty) intention. In this case, it is necessary to further instantiate the plan below the finished plan at the top of that intention, and remove the goal that was left at the beginning of the body of the plan below (see rules **AchvGl** and **TestGl$_2$**). Note that, in this case, further 'clearing' might be necessary, hence the next step is still **ClrInt**. Rule **ClrInt$_3$** takes care of the situation where no (further) clearing is required, so a new reasoning cycle can start (step **ProcMsg**).

$$\frac{T_\iota = [head \leftarrow \top]}{\langle ag, C, M, T, \mathsf{ClrInt} \rangle \longrightarrow \langle ag, C', M, T, \mathsf{ProcMsg} \rangle} \quad \text{(ClrInt}_1\text{)}$$

$$\text{where:} \quad C'_I = C_I \setminus \{T_\iota\}$$

$$\frac{T_\iota = i[head \leftarrow \top]}{\langle ag, C, M, T, \mathsf{ClrInt} \rangle \longrightarrow \langle ag, C', M, T, \mathsf{ClrInt} \rangle} \quad \text{(ClrInt}_2\text{)}$$

$$\text{where:} \quad C'_I = (C_I \setminus \{T_\iota\}) \cup \{k[(head' \leftarrow h)\theta]\}$$
$$\text{if } i = k[head' \leftarrow g; h] \text{ and } \theta \text{ is s.t. } g\theta = \mathsf{TrEv}(head)$$

$$\frac{T_\iota \neq [head \leftarrow \top] \wedge T_\iota \neq i[head \leftarrow \top]}{\langle ag, C, M, T, \mathsf{ClrInt} \rangle \longrightarrow \langle ag, C, M, T, \mathsf{ProcMsg} \rangle} \quad \text{(ClrInt}_3\text{)}$$

10.2 Semantics of Message Exchange in a Multi-Agent System

The format of messages is $\langle mid, id, ilf, cnt \rangle$, where mid uniquely identifies the message (or message exchange), id identifies the agent to which the message is addressed (when the message is being sent) or the agent that has sent the message (when the message is being received), ilf is the illocutionary force (performative) associated with the message, and cnt is the message content, which can be: an atomic formula (*at*), a set of formulæ (*ATs*), a ground atomic formula (*b*), a set of ground atomic formulæ (*Bs*) or a set of plans (*PLs*), depending on the illocutionary force associated with the message.

A mechanism for receiving and sending messages asynchronously is then defined. Messages are stored in a mail box and one of them is processed by the agent at the beginning of a reasoning cycle. Recall that, within a configuration of the semantics, M_{In} is the set of messages that the agent has received but has not processed yet, M_{Out} is the set of messages to be sent to other agents, and M_{SI} is a set of suspended intentions awaiting replies of (information request) messages previously sent. More specifically, M_{SI} is a set of pairs of the form (*mid*, *i*), where *mid* is a message identifier that uniquely identifies the previously sent message that caused intention *i* to be suspended.

When sending messages with illocutionary forces related to information request, we have opted for a semantics in which the intention is suspended (see rule **ExecActSndAsk** below) until a reply is received from the interlocutor, very much in the way in which intentions get suspended when they are waiting for an internal event to be selected and handled. With this particular semantics for 'ask' messages, the programmer knows for a fact that any subsequent action in the body of a plan is only executed after the requested information has already been received – however, note that, differently from *Jason*, we here assume that the information retrieved is stored directly in the agent's belief base, so a test goal is required if the information is to be used in the remainder of the plan.

We now give two rules for executing the (internal) action of sending a message to another agent. These rules have priority over **Action**; although **Action** could also be applied on the same configurations, we assume the rules below are used if the formula to be executed is specifically a .send. We did not include this as a proviso in rule **Action** to increase readability.

$$\frac{\begin{array}{c} T_\iota = i[head \leftarrow .\mathtt{send}(id, ilf, cnt); h] \\ ilf \in \{AskOne, AskAll, AskHow\} \end{array}}{\langle ag, C, M, T, \mathsf{ExecInt} \rangle \longrightarrow \langle ag, C', M', T, \mathsf{ProcMsg} \rangle} \; \textbf{(ExecActSndAsk)}$$

$$\begin{aligned} \text{where:} \quad M'_{Out} &= M_{Out} \cup \{\langle mid, id, ilf, cnt \rangle\} \\ M'_{SI} &= M_{SI} \cup \{(mid, i[head \leftarrow h])\}, \\ &\quad \text{with } mid \text{ a new message identifier;} \\ C'_I &= (C_I \setminus \{T_\iota\}) \end{aligned}$$

The semantics of sending other types of illocutionary forces is then simply to add a well-formed message to the agent's mail outbox (rule **ExecActSnd** below). Note that in the rule above, as the intention is suspended, the next step in the reasoning cycle is ProcMsg (i.e. a new cycle is started), whereas in the rule below it is ClrInt, as the updated intention – with the sending action removed from the plan body – might require 'clearing', as with most of the intention execution rules seen in the previous section.

$$T_\iota = i[head \leftarrow .send(id, ilf, cnt); h]$$
$$ilf \notin \{AskOne, AskAll, AskHow\}$$

$$\overline{\langle ag, C, M, T, \mathsf{ExecInt} \rangle \longrightarrow \langle ag, C', M', T, \mathsf{ClrInt} \rangle}$$ (ExecActSnd)

where: $M'_{Out} = M_{Out} \cup \{\langle mid, id, ilf, cnt \rangle\}$,
                 with mid a new message identifier;
         $C'_I = (C_I \setminus \{T_\iota\}) \cup \{i[head \leftarrow h]\}$

Whenever new messages are sent, we assume the system creates new *message identifiers* (*mid*) which are unique in the whole system. Later we shall see that, when replying to a message, the same message identifier is kept in the message, similar to the way `reply-with` is used in KQML. Thus the receiving agent is aware that a particular message is a reply to a previous one by checking the message identifiers in the set of intentions that were suspended waiting for a reply. This feature will be used when we give semantics to receiving *Tell* messages, which can be used by an agent when it spontaneously wants the receiver to believe something (or at least to believe something about the sender's beliefs), but can also be used when the agent receives an 'ask' type of message and chooses to reply to it.

The semantics here is not intended to give the whole picture of an agent architecture and a multi-agent system, but as we will present the semantics of communication, it helps to show a simple rule that gives the notion of message exchange. In the same way that an overall agent architecture includes a specific architecture for its reasoner (in this case an AgentSpeak interpreter), the sensors and the effectors, it is also very usual to assume that the architecture includes the means for agents to send messages to other agents. This is abstracted away in the semantics by means of the following rule, where each $AG_{id_k}, k = 1 \ldots n$, is an agent configuration $\langle ag_{id_k}, C_{id_k}, M_{id_k}, T_{id_k}, s_{id_k} \rangle$:

$$\langle mid, id_j, ilf, cnt \rangle \in M_{id_i\,Out}$$

$$\overline{\{AG_{id_1}, \ldots AG_{id_i}, AG_{id_j}, \ldots AG_{id_n}, env\} \longrightarrow}$$ (MsgExchg)
$$\{AG_{id_1}, \ldots AG'_{id_i}, AG'_{id_j}, \ldots AG_{id_n}, env\}$$

where: $M'_{id_i\,Out} = M_{id_i\,Out} \setminus \{\langle mid, id_j, ilf, cnt \rangle\}$
         $M'_{id_j\,In} = M_{id_j\,In} \cup \{\langle mid, id_i, ilf, cnt \rangle\}$

In the rule above, there are n agents in the society and *env* denotes the environment in which the agents are situated; typically, this is not an AgentSpeak agent, it is simply represented as a set of properties currently true in the environment and how they are changed by an agent's actions. Note how, in a message that is to be sent, the second component identifies the addressee (the agent to which the message is being sent), whereas in a received message that same component identifies the sender of the message.

10.3 Semantic Rules for Receiving Messages

In Section 6.2, we saw that the KQML-like performatives were implemented as a set of AgentSpeak plans that can be redefined in an agent source code. We here give a formal account of what those predefined plans do.

For processing messages, a new selection function is necessary, which operates in much the same way as the three selection functions described in the previous section. The new selection function is called $\mathcal{S}_{\mathcal{M}}$, and selects one particular message from M_{In}; intuitively, it represents the priority assigned to each type of message by the programmer. We also need another 'given' function, in the sense that it is assumed to be given in an agent's specification (as selection functions are), but its purpose is different from selection functions. The Boolean function SocAcc(id, ilf, at), where ilf is the illocutionary force of the message from agent id, with propositional content at, determines when a message is *socially acceptable* in a given context. For example, for a message of the form $\langle mid, id, Tell, at \rangle$, the receiving agent may want to consider whether id is a sufficiently trustworthy source of information so that even registering the fact that such agent believes at to be the case is appropriate or not. For a message with illocutionary force *Achieve*, the agent could check, for example, whether id has sufficient social power over itself, or whether it wishes to act altruistically towards id and then do whatever it is being asked. The $\mathcal{S}_{\mathcal{M}}$ and SocAcc functions can be defined by the user as explained in Section 7.2.

The annotation mechanism introduced earlier provides a useful notation for making explicit the sources of an agent's belief. It has advantages in terms of readability, besides simplifying the use of such explicit information in an agent's reasoning. Note that, with this language construct, it is straightforward to determine, in a plan context, what the source of a belief was before using that plan as an intended means. Before we start the presentation of the semantic rules for communication, we remark that the semantics does not consider *nested* annotations (i.e. to allow the representation of situations in which i was told φ by j, who in turn was told φ by k, and so on).

Receiving a *Tell* message: a *Tell* message might be sent to an agent either as a reply or as an inform action. When receiving a *Tell* message as an inform (as opposed to a reply to a previous information request), the AgentSpeak agent will include the content of the received message in its knowledge base and will annotate the sender as a source for that belief. Note that this corresponds, in a way, to what is specified as 'action completion' by Labrou and Finin [61]: the receiver will know about the sender's attitude regarding that belief. To account for agent autonomy and the social aspects of multi-agent systems, we consider that social relations will regulate which messages the receiver will indeed process; this is referred to in the semantics by the SocAcc function.

$$\mathcal{S}_{\mathcal{M}}(M_{In}) = \langle mid, id, Tell, Bs \rangle$$
$$(mid, i) \notin M_{SI} \text{ (for any intention } i)$$
$$SocAcc(id, Tell, Bs)$$

$$\overline{\langle ag, C, M, T, \mathsf{ProcMsg} \rangle \longrightarrow \langle ag', C', M', T, \mathsf{SelEv} \rangle} \quad \textbf{(Tell)}$$

$$\text{where:} \quad M'_{In} \; = \; M_{In} \setminus \{\langle mid, id, Tell, Bs \rangle\}$$

and for each $b \in Bs$:
$$ag'_{bs} \; = \; ag_{bs} + b[id]$$
$$C'_E \; = \; C_E \cup \{\langle +b[id], \top \rangle\}$$

Receiving a *Tell* message as a reply: this rule is similar to the one above, except that now the suspended intention associated with that particular message, given that it is a reply to a previous request from this agent, needs to be activated again.

$$\mathcal{S}_{\mathcal{M}}(M_{In}) = \langle mid, id, Tell, Bs \rangle$$
$$(mid, i) \in M_{SI} \text{ (for some intention } i)$$
$$SocAcc(id, Tell, Bs)$$

$$\overline{\langle ag, C, M, T, \mathsf{ProcMsg} \rangle \longrightarrow \langle ag', C', M', T, \mathsf{SelEv} \rangle} \quad \textbf{(TellRepl)}$$

$$\text{where:} \quad M'_{In} \; = \; M_{In} \setminus \{\langle mid, id, Tell, Bs \rangle\}$$
$$M'_{SI} \; = \; M_{SI} \setminus \{(mid, i)\}$$
$$C'_I \; = \; C_I \cup \{i\}$$

and for each $b \in Bs$:
$$ag'_{bs} \; = \; ag_{bs} + b[id]$$
$$C'_E \; = \; C_E \cup \{\langle +b[id], \top \rangle\}$$

We should emphasise at this point that the semantics for receiving *Tell* and *Untell* messages as a reply to a previous question makes simplifying assumptions and therefore does not match all the details of the ***Jason*** implementation. In ***Jason***, the wait for a reply can time out, and there are two ways of asking questions: one in which the intention is suspended and the reply is instantiated in an extra parameter of the .send action when the reply is received (but not added to the belief base), and another in which the plan execution can carry on after the asking message is sent to the other agent, and the reply then only affects the belief base. For details, refer to Section 6.2.

Receiving an *Untell* message: when receiving an *Untell* message, the sender of the message is removed from the set of sources giving accreditation to the atomic formula in the content of the message. In cases where the sender was the only source for that information, the belief itself is removed from the receiver's belief base. Note that, as the atomic formula in the content of an *Untell* message can have uninstantiated variables, each belief in the agent's belief base that can be unified with that formula needs to be considered in turn.

$$\frac{\begin{array}{c} \mathcal{S}_{\mathcal{M}}(M_{In}) = \langle mid, id, Untell, ATs \rangle \\ (mid, i) \notin M_{SI} \text{ (for some intention } i) \\ \mathsf{SocAcc}(id, Untell, ATs) \end{array}}{\langle ag, C, M, T, \mathsf{ProcMsg} \rangle \longrightarrow \langle ag', C', M', T, \mathsf{SelEv} \rangle} \quad \text{(Untell)}$$

$$\begin{aligned} \text{where:} \quad M'_{In} \;&=\; M_{In} \setminus \{\langle mid, id, Untell, ATs \rangle\} \\[4pt] &\text{and for each } b \in \{at\theta \mid \\ &\qquad \theta \in \mathsf{Test}(ag_{bs}, at) \land at \in ATs\} \\ ag'_{bs} \;&=\; ag_{bs} - b[id] \\ C'_E \;&=\; C_E \cup \{\langle -b[id], \top \rangle\} \end{aligned}$$

Receiving an *Untell* message as a reply: as above, the sender as source for the belief, or the belief itself, is excluded from the belief base of the receiver, except that now a suspended intention needs to be resumed.

$$\frac{\begin{array}{c} \mathcal{S}_{\mathcal{M}}(M_{In}) = \langle mid, id, Untell, ATs \rangle \\ (mid, i) \in M_{SI} \text{ (for some intention } i) \\ \mathsf{SocAcc}(id, Untell, ATs) \end{array}}{\langle ag, C, M, T, \mathsf{ProcMsg} \rangle \longrightarrow \langle ag', C', M', T, \mathsf{SelEv} \rangle} \quad \text{(UntellRepl)}$$

$$\begin{aligned} \text{where:} \quad M'_{In} \;&=\; M_{In} \setminus \{\langle mid, id, Untell, ATs \rangle\} \\ M'_{SI} \;&=\; M_{SI} \setminus \{(mid, i)\} \\ C'_I \;&=\; C_I \cup \{i\} \\[4pt] &\text{and for each } b \in \{at\theta \mid \\ &\qquad \theta \in \mathsf{Test}(ag_{bs}, at) \land at \in ATs\} \\ ag'_{bs} \;&=\; ag_{bs} - b[id] \\ C'_E \;&=\; C_E \cup \{\langle -b[id], \top \rangle\} \end{aligned}$$

Receiving an *Achieve* message: in an acceptable social context (e.g. if the sender has 'power' over the receiver), the receiver will try to execute a plan whose triggering event is $+!at$; that is, it will try to achieve the goal associated with the propositional content of the message. An external event is thus included in the set of events (recall that external events have the triggering event associated with the empty intention ⊤).

Note that it is now possible to have a new focus of attention (the stacks of plans in the set of intentions I) being started by the addition (or deletion, see below) of an achievement goal. Originally, only a change of belief from perception of the environment started a new focus of attention; the plan chosen for that event could, in turn, have achievement goals in its body, thus pushing new plans onto the stack.

$$\frac{\mathcal{S}_\mathcal{M}(M_{In}) = \langle mid, id, Achieve, at \rangle \\ \mathsf{SocAcc}(id, Achieve, at)}{\langle ag, C, M, T, \mathsf{ProcMsg} \rangle \longrightarrow \langle ag, C', M', T, \mathsf{SelEv} \rangle} \quad \text{(Achieve)}$$

$$\text{where:} \quad \begin{aligned} M'_{In} &= M_{In} \setminus \{\langle mid, id, Achieve, at \rangle\} \\ C'_E &= C_E \cup \{\langle +!at, \top \rangle\} \end{aligned}$$

Receiving an *Unachieve* message: the way this has been defined in the formal semantics is not the same as we have chosen to implement in *Jason*. When an *Unachieve* message is received, the desires and intentions to achieve that goal are simply dropped (note that this only happens if the message is accepted, of course). This is done using the `.drop_desire` standard internal action (see Appendix A); formal semantics for this will only be available after those internal actions themselves have been given formal semantics, which remains ongoing work.

Receiving a *TellHow* message: the notion of plan in reactive planning systems such as those defined by AgentSpeak agents is associated with Singh's notion of know-how [89]. Accordingly, we use the *TellHow* performative when an external source wants to inform an AgentSpeak agent of a plan it uses for handling certain types of events (given by the plan's triggering event). If the source is trusted, the plans (which are in the message content) are simply added to the agent's plan library.

$$\frac{\mathcal{S}_\mathcal{M}(M_{In}) = \langle mid, id, TellHow, PLs \rangle \\ (mid, i) \notin M_{SI} \text{ (for any intention } i) \\ \mathsf{SocAcc}(id, TellHow, PLs)}{\langle ag, C, M, T, \mathsf{ProcMsg} \rangle \longrightarrow \langle ag', C, M', T, \mathsf{SelEv} \rangle} \quad \text{(TellHow)}$$

$$\text{where:} \quad \begin{aligned} M'_{In} &= M_{In} \setminus \{\langle mid, id, TellHow, PLs \rangle\} \\ ag'_{ps} &= ag_{ps} \cup PLs \end{aligned}$$

Note that we do not include an annotation to explicitly state what was the source of that plan; that is, with the semantics as defined here, the agent cannot reason about the agent that provided such know-how when making a decision on whether to use that plan at a later point in time. However, recall that this is already available in the *Jason* interpreter, as the language was extended with the use of annotated predicates as plan labels.

Receiving a *TellHow* message as a reply: the *TellHow* performative as a reply will cause the suspended intention associated with that particular message exchange to be resumed.

$$\frac{\begin{array}{c} \mathcal{S}_\mathcal{M}(M_{In}) = \langle mid, id, TellHow, PLs \rangle \\ (mid, i) \in M_{SI} \text{ (for some intention } i) \\ \mathsf{SocAcc}(id, TellHow, PLs) \end{array}}{\langle ag, C, M, T, \mathsf{ProcMsg} \rangle \longrightarrow \langle ag', C', M', T, \mathsf{SelEv} \rangle} \quad \text{(TellHowRepl)}$$

where:
$$\begin{array}{rcl} M'_{In} & = & M_{In} \setminus \{\langle mid, id, TellHow, PLs \rangle\} \\ M'_{SI} & = & M_{SI} \setminus \{(mid, i)\} \\ C'_I & = & C_I \cup \{i\} \\ ag'_{ps} & = & ag_{ps} \cup PLs \end{array}$$

Receiving an *UntellHow* message: this is similar to rule **TellHow**. An external source may find that a plan is no longer appropriate for handling the events it was supposed to handle; it may then want to inform that to another agent. Thus, when receiving a socially acceptable *UntellHow* message, the agent removes the associated plans (in the message content) from its plan library.

$$\frac{\begin{array}{c} \mathcal{S}_\mathcal{M}(M_{In}) = \langle mid, id, UntellHow, PLs \rangle \\ \mathsf{SocAcc}(id, UntellHow, PLs) \end{array}}{\langle ag, C, M, T, \mathsf{ProcMsg} \rangle \longrightarrow \langle ag', C, M', T, \mathsf{SelEv} \rangle} \quad \text{(UntellHow)}$$

where:
$$\begin{array}{rcl} M'_{In} & = & M_{In} \setminus \{\langle mid, id, UntellHow, PLs \rangle\} \\ ag'_{ps} & = & ag_{ps} \setminus PLs \end{array}$$

Receiving an *AskOne* message: the receiver will respond to the request for information (we consider that asking is a special request) if certain conditions imposed by the social settings (the SocAcc function) hold between sender and receiver.

Note that *AskOne* and *AskAll* differ essentially in the kind of request made to the receiver. With the former, the receiver should just confirm whether the received predicate (in the message content) is in its belief base or not; with the latter, the agent replies with all the predicates in the belief base that unify with the formula in the message content. The receiver processing an *AskOne* message has to respond with the action of sending either a *Tell* or *Untell*, provided that the social configuration is such that the receiver will consider the sender's request.

$$\frac{\mathcal{S}_{\mathcal{M}}(M_{In}) = \langle mid, id, AskOne, \{b\}\rangle \quad \mathsf{SocAcc}(id, AskOne, b)}{\langle ag, C, M, T, \mathsf{ProcMsg}\rangle \longrightarrow \langle ag, C, M', T, \mathsf{SelEv}\rangle}$$

where:

$$M'_{In} = M_{In} \setminus \{\langle mid, id, AskOne, \{b\}\rangle\}$$

$$M'_{Out} = \begin{cases} M_{Out} \cup \{\langle mid, id, Tell, AT\rangle\}, \\ \quad AT = \{at\theta\}, \ \theta \in \mathsf{Test}(ag_{bs}, at) & \text{if } \mathsf{Test}(ag_{bs}, at) \neq \{\} \\ M_{Out} \cup \{\langle mid, id, Untell, \{at\}\rangle\} & \text{otherwise} \end{cases}$$

(AskOne)

The role that $\mathcal{S}_{\mathcal{M}}$ plays in the agent's reasoning cycle is now slightly more important than originally conceived [70]. An agent considers whether to accept a message or not, but the reply message is automatically assembled when the agent selects any of the 'ask' types of messages. However, providing such a reply may require considerable computational resources (e.g. a large plan library may need to be scanned and a considerable number of plans retrieved from it in order to produce a reply message). Therefore, $\mathcal{S}_{\mathcal{M}}$ should normally be defined so that the agent only selects an *AskOne*, *AskAll* or *AskHow* message if it determines it is not currently too busy to provide a reply.

Receiving an *AskAll*: as for *AskOne*, the receiver processing an *AskAll* has to respond either with *Tell* or *Untell*, provided the social context is such that the receiver will accept such request from the sender. As mentioned before, here the agent replies with all the predicates in the belief base that unify with the formula in the message content.

$$\frac{\mathcal{S}_{\mathcal{M}}(M_{In}) = \langle mid, id, AskAll, \{at\}\rangle \quad \mathsf{SocAcc}(id, AskAll, at)}{\langle ag, C, M, T, \mathsf{ProcMsg}\rangle \longrightarrow \langle ag, C, M', T, \mathsf{SelEv}\rangle}$$

where:

$$M'_{In} = M_{In} \setminus \{\langle mid, id, AskAll, \{at\}\rangle\}$$

$$M'_{Out} = \begin{cases} M_{Out} \cup \{\langle mid, id, Tell, ATs\rangle\}, \\ \quad ATs = \{at\theta \mid \theta \in \mathsf{Test}(ag_{bs}, at)\} & \text{if } \mathsf{Test}(ag_{bs}, at) \neq \{\} \\ M_{Out} \cup \{\langle mid, id, Untell, \{at\}\rangle\} & \text{otherwise} \end{cases}$$

(AskAll)

Receiving an *AskHow*: the receiver processing an *AskHow* has to respond with the action of sending a *TellHow*, provided the social configuration is such that the receiver will consider the sender's request. In contrast to the use of *Untell* in

AskAll, the response when the receiver knows no relevant plan is a reply with an empty set of plans.

$$\frac{\begin{array}{c} \mathcal{S}_{\mathcal{M}}(M_{In}) = \langle mid, id, AskHow, te \rangle \\ \mathsf{SocAcc}(id, AskHow, te) \end{array}}{\langle ag, C, M, T, \mathsf{ProcMsg} \rangle \longrightarrow \langle ag, C, M', T, \mathsf{SelEv} \rangle} \quad \textbf{(AskHow)}$$

where:
$$\begin{array}{rcl} M'_{In} & = & M_{In} \setminus \{\langle mid, id, AskHow, te \rangle\} \\ M'_{Out} & = & M_{Out} \cup \{\langle mid, id, TellHow, PLs \rangle\} \\ & & \text{and } PLs = \{p \mid (p, \theta) \in \mathsf{RelPlans}(ag_{ps}, te)\} \end{array}$$

When SocAcc fails: all the rules above consider the social relations holding for sender and receiver (i.e. they require SocAcc to be true). If the required social relation does not hold, the message is simply removed from the set of messages. The rule below is used for receiving a message from an untrusted source, regardless of the performative.

$$\frac{\begin{array}{c} \mathcal{S}_{\mathcal{M}}(M_{In}) = \langle mid, id, ilf, Bs \rangle \\ \neg\mathsf{SocAcc}(id, ilf, Bs) \\ (\text{with } ilf \in \{Tell, Untell, TellHow, UntellHow, \\ Achieve, Unachieve, AskOne, AskAll, AskHow\}) \end{array}}{\langle ag, C, M, T, \mathsf{ProcMsg} \rangle \longrightarrow \langle ag, C, M', T, \mathsf{SelEv} \rangle} \quad \textbf{(NotSocAcc)}$$

$$\text{where:} \quad M'_{In} \quad = \quad M_{In} \setminus \{\langle mid, id, ilf, Bs \rangle\}$$

When M_{In} is empty: this last semantic rule states that, when the mail inbox is empty, the agent simply goes to the next step of the reasoning cycle (SelEv).

$$\frac{M_{In} = \{\}}{\langle ag, C, M, T, \mathsf{ProcMsg} \rangle \longrightarrow \langle ag, C, M, T, \mathsf{SelEv} \rangle} \quad \textbf{(NoMsg)}$$

10.4 Semantics of the BDI Modalities for AgentSpeak

In [10], a way of interpreting the informational, motivational and deliberative modalities of BDI logics for AgentSpeak agents was given, based on the operational semantics. We use that same framework for interpreting the BDI modalities when we implemented the standard internal actions which allows an agent to check its own desires and intentions (see Appendix A.3).

We here show only the main definitions given in [10], without covering much of the motivation for our choices. In particular, discussion is given in that paper on the interpretation of intentions and desires, as the *belief* modality is clearly defined in AgentSpeak. Although the configuration of the transition system is as defined earlier (i.e. $\langle ag, C, M, T, s \rangle$), the definitions below make use only of the *ag* and *C* components, so we refer to a configuration being simply $\langle ag, C \rangle$ to make the presentation clearer.

Definition 10.5 (Belief) *We say that an AgentSpeak agent ag, regardless of its circumstance C, believes a formula φ iff it is included in the agent's belief base; that is, for an agent $ag = \langle bs, ps \rangle$:*

$$\mathrm{BEL}_{\langle ag, C \rangle}(\varphi) \quad \equiv \quad bs \models \varphi.$$

Note that closed world assumption is used here, so $\mathrm{BEL}_{\langle ag, C \rangle}(\varphi)$ is true if φ is included in the agent's belief base, and $\mathrm{BEL}_{\langle ag, C \rangle}(\neg\varphi)$ is true otherwise, where φ is an atom (i.e. *at* in the grammar given in page 201).

Before giving the formal definition for the *intention* modality, we first define an auxiliary function $agls : \mathcal{I} \to \mathcal{P}(\Phi)$, where $\mathcal{I}$ is the domain of all individual intentions and Φ is the domain of all atomic formulæ (as mentioned above). Recall that an intention is a stack of partially instantiated plans, so the definition of $\mathcal{I}$ is as follows. The empty intention (or true intention) is denoted by $\top$, and $\top \in \mathcal{I}$. If p is a plan and $i \in \mathcal{I}$, then also $i[p] \in \mathcal{I}$. The *agls* function below takes an intention and returns all achievement goals in the triggering event part of the plans in it:

$$
\begin{aligned}
agls(\top) &= \{\} \\
agls(i[p]) &= \begin{cases} \{at\} \cup agls(i) & \text{if } p = +!at : ct \leftarrow h \\ agls(i) & \text{otherwise.} \end{cases}
\end{aligned}
$$

Definition 10.6 (Intention) *We say an AgentSpeak agent ag intends φ in circumstance C if, and only if, it has φ as an achievement goal that currently appears in its set of intentions C_I, or φ is an achievement goal that appears in the (suspended) intentions[2] associated with events in C_E. For an agent ag and circumstance C, we have:*

$$\mathrm{INTEND}_{\langle ag, C \rangle}(\varphi) \equiv \varphi \in \bigcup_{i \in C_I} agls(i) \ \lor \ \varphi \in \bigcup_{\langle te, i \rangle \in C_E} agls(i).$$

Note that we are only interested in atomic formulæ *at* in triggering events that have the form of additions of achievement goals, and ignore all other types of triggering events. These are the formulæ that represent (symbolically) properties

---

[2]Note that here we have a simplified notion of suspended intentions. In **Jason**, there are other structures used to keep all the suspended intentions, whether they are suspended waiting for a reply from an 'ask' type of message or waiting for an action execution, besides waiting for an internal event to be handled.

of the states of the world that the agent is trying to achieve (i.e. the intended states). However, taking such formulæ from the agent's set of intentions does not suffice for defining intentions, as there can be *suspended* intentions. Suspended intentions are precisely those that appear in the set of events.

We are now in a position to define the interpretation of the *desire* modality in AgentSpeak agents.

Definition 10.7 (Desire) *We say that an agent in circumstance C desires a formula φ if, and only if, φ is an achievement goal in C's set of events C_E (associated with any intention i), or φ is a current intention of the agent; more formally:*

$$\text{DES}_{\langle ag,C \rangle}(\varphi) \equiv \langle +!\varphi, i \rangle \in C_E \ \lor \ \text{INTEND}_{\langle ag,C \rangle}(\varphi).$$

Although this is not discussed in the original literature on AgentSpeak, it was argued in [10] that the *desire* modality in an AgentSpeak agent is best represented by additions of achievement goals presently in the set of events, as well as its present intentions.

The code in Section 9.1 shows various examples of the use of the *Jason* internal actions which implement the BDI modalities as defined above, which can be useful for checking whether an agent currently has certain desires or intentions. While it is an integral part of the AgentSpeak language to check, in the context of a plan, whether an agent has certain beliefs, the other modalities require special internal actions. Those internal actions are informally described in Appendix A.3. Because these actions, as well as other internal actions such as those used to drop desires and intentions, make explicit reference to (and even change) the structures used in the formal semantics, they too should be formalised. As mentioned earlier, the formal semantics for the BDI standard internal actions available in *Jason* is the subject of ongoing research.

11

Conclusions

We hope this book has given a comprehensive overview of all that is involved in developing multi-agent systems using *Jason*. The *Jason* platform has evolved a great deal over the last few years, thanks partly to its user base. In this chapter we will discuss various issues related to *Jason*, its use and its future directions.

11.1 *Jason* and Agent-Oriented Programming

Recent years have seen a rapid increase in the amount of research being done on agent-oriented programming languages, and multi-agent systems techniques that can be used in the context of an agent programming language. The number of different programming language, tools and platforms for multi-agent systems that has appeared in the literature [12] really is impressive, in particular logic-based languages [47]. In [14], some of the languages that have working interpreters of practical use were presented in reasonable detail. It is not our aim here to discuss or even list all the existing agent languages and platforms, but some examples are 3APL [35] (and its recent 2APL variation), MetateM [46], ConGolog [37], CLAIM [43], IMPACT [42], Jadex [78], JADE [9], SPARK [72], MINERVA [64], SOCS [1, 97], Go! [26], STEAM [94], STAPLE [60], JACK [101]. Other languages have been discussed in the survey papers mentioned above (i.e. [12, 47]) and also in the surveys available in [68] and [32].

An important current trend in the literature on programming languages for multi-agent systems is the idea of declarative goals and reasoning about declarative goals [23, 55, 83, 95, 96, 98, 102]. In most cases, what is done is that the semantics of the language is extended so that interpreters for the language also maintain a *goal base*. Because of the elegant simplicity of the AgentSpeak language, much

Programming Multi-Agent Systems in AgentSpeak using Jason   R.H. Bordini, J.F. Hübner, M. Wooldridge
© 2007 John Wiley & Sons, Ltd

misconception about it has been popularised by word of mouth and sometimes even in the literature.

The first misconception is that AgentSpeak 'reduces' the notion of a goal to that event. The fact that *additions of new goals* and *dropping a plan to achieve a goal* generate (internal) events, in the same way that addition/deletion of beliefs due to perception of the environment generates (external) events, is in fact a very important advantage of the language. This allows the programmer to develop domain-specific event selection functions which, for that particular application, can balance the agent's reactive and pro-active behaviour, as both types of events are 'competing' to be selected at the beginning of each (practical) reasoning cycle. This by no means obscures the fact that the language has a very clear/specific construct to express *goals* (in the 'desire' sense rather than the Prolog sense, for those familiar with the logic programming parlance). What AgentSpeak does not do is to *enforce* a declarative use of the goal construct; however, as pointed out in [99], programmers can do this easily by ensuring that a plan to achieve a goal g only finishes successfully if g is believed by the agent at the end of the plan's execution: there are predefined patterns to help programmers do that in *Jason*, as discussed in Chapter 8. Also, as other languages include a 'perform goal' construct, it is generally agreed that it is important to have the possibility of using plans for often repeated (minor) tasks which do not necessarily result in new beliefs but also need further consideration on the most appropriate way to execute them, and need doing so as late as possible given the dynamic environments that agents typically share.

Another misconception is that 'AgentSpeak does not have a goal base'. Much before other languages realised they needed such a thing, AgentSpeak already had a much more sophisticated structure which subsumes a goal base: the set of intentions (which we sometimes call the 'intentional structure'). Such intentional structure not only records all the *goals* that an agent is currently pursuing but also the particular *means* that it committed itself to use in order to achieve those goals. With that structure (which, recall, is a set of stacks of partially instantiated plans), it is also possible to retrieve other information such as the known goals that the agent will come to have in the future, and – with the use of *Jason*'s extensibility features – we can also retrieve plans that failed to achieve the respective goals, as well as the contingency plans that are being used in order to attempt again to achieve those goals or at least to 'clean up' a possible messy state obtained because of a partly executed plan that failed.

The reason why *Jason* was developed from the abstract AgentSpeak(L) language originally defined by Rao [80] is precisely the elegance of its notation and the coherence with which it involves all the most important elements of the BDI architecture, which in many different ways and guises has influenced a significant part of the research in the area of autonomous agents and multi-agent systems.

Also, because *Jason* has a logic-based programming language but is easily extensible using Java, it is straightforward to combine it with major trends in software development (in particular for the Web) such as ontologies [91] and web services [88]. The case study in Chapter 9.2 showed how a web application can be developed with *Jason* using JSP, and the 'amazon client' application available in the *Jason* download page shows how web services can be easily accessed by an agent. One of the research strands involving *Jason* mentioned below is related to ongoing work to use ontologies as part of a *Jason* agent's belief base.

11.2   Ongoing Work and Related Research

There are also some constructs discussed in the agent-oriented programming literature [12, 14] which are not, or not yet, directly available in *Jason*. One such construct is called *conjunctive goals* [99]. Typically an agent has a goal of achieving a state of the environment where one particular property of the environment becomes true. However, sometimes it facilitates modelling if we can say that the goal is to achieve two (or more) properties at the same time, and that is a conjunctive goal. For the time being, we can (partially) handle such conjunctive goals with 'tricks' (which could easily be developed as a pattern) such as:

```
// belief rules
...
conj(P,Q) :- P & Q.

//plans
...
<some_event> : <some_context>
  <- !conj_goal(a,b). // needs to achieve (a & b), i.e.
                      // a and b being simultaneously true

@plan_conj_goal // you may consider using annotation "[atomic]" here
+!conj_goal(G1,G2) <- !G1; !G2; ?conj(G1,G2).

// you will probably need contingency plans here, depending
// on the commitment strategy you need for your conjunctive goals.
// you can also replace "!G1; !G2" with a known course of action that
// is expected to bring about both G1 and G2 simultaneously.
```

Another example of a construct not available in *Jason* is to allow *parallel goals* as available, e.g. in CAN [102]. This allows us to achieve two goals concurrently and, when both are achieved, carry on with a given course of action. *Jason* provides the !! operator which can be used to create a separate intention, which then becomes completely independent, and recall that the agent can interleave the execution of its various intentions. We avoided adding the parallel construct as we feared it might complicate things more than help, as programmers already have to manage the fact that various intentions compete for the agent's focus of attention, so having further concurrency within a plan might make things too difficult to manage. On the other hand, there might be applications where this would be natural to program. We are still pondering in order to get the right balance between having enough and too many language constructs, so we might still decide to add parallel goals in future *Jason* releases.

The agent-oriented programming literature has been discussing profusely the notion of *capabilities* available in languages such as [78, 101] and methodologies such as [75]. Essentially, capabilities encapsulate sets of plans aimed at particular (sets of) tasks in such a way that they do not interfere with other plans in other capabilities; clearly, this facilitates reuse of plans in agent development. Although this feature is not available in *Jason* 1.0 (the release described in this book), it will be available within the next few releases. Another possibility is to consider the use of the notion of roles as an encapsulation mechanism for agent languages, as put forward in [34].

From the above discussion it should be clear that *Jason* is still, as is indeed the whole area of agent-oriented programming, very much work in progress. Besides those relatively straightforward language extensions just discussed, there are various strands of research that make direct use of *Jason*, and which are likely to result in techniques that may greatly facilitate the use of *Jason* for the development of sophisticated multi-agent systems. We describe some of these below.

> **Organisations:** an important part of agent-oriented software engineering is related to agent *organisations*, which has received much research attention in the last few years. We are currently working on allowing specifications of agent organisations (with the related notions of roles, groups, relationships between groups, social norms, etc.) to be used in combination with *Jason* for programming the individual agents. The particular organisational model we use is $\mathcal{M}oise^+$ [56] and its combination with *Jason* is available at http://moise.sf.net.

> **Plan exchange:** work has been done to allow plan exchange between AgentSpeak agents, which can be very useful, in particular for systems of cooperating agents, but also for applications in which a large number of plans cannot be kept in the agent's plan library simultaneously (e.g. for use in PDAs with limited computational resources). The motivation for this work is this simple intuition: if you do not know how to do something, ask someone who does. However, various issues need to be considered in engineering systems where such plan exchanges can happen (e.g. which plans can be exchanged, what to do with a plan

retrieved from another agent, who and when to ask for plans). This work is based on the Coo-BDI plan exchange mechanism [5].

Ontological reasoning: although this is not available in *Jason* yet, in [71] it was argued that the belief base of an AgentSpeak agent should be formulated as a (populated) ontology, whereby:

1. Queries to the belief base are more expressive as their results do not rely only on explicitly represented literals but can be inferred from the ontology.

2. Retrieving a plan for handling an event is more flexible as it is not based solely on unification but also on the subsumption relation between concepts.

3. Agents may share knowledge by using ontology languages such as OWL.

4. The notion of belief update can be refined given that the (ontological) consistency of a belief addition can be checked.

Concretely, in *Jason* we plan to use annotations to specify which ontology each belief belongs to, and to use an existing tool to do the ontological reasoning when required. This further increases the need to have appropriate belief revision.

Belief revision: in *Jason* (and most other agent-oriented programming platforms) consistency of the belief base is left for programmers to ensure. The reason is that automatic belief revision is typically too expensive computationally, and thus not appropriate for practical agent programming. In [3], a new (polynomial-time) algorithm for belief revision was introduced. The algorithm is tractable due to simplifying assumptions which are nevertheless realistic for belief bases as used in agent programming. We plan to make available an implementation of an adaptation of that algorithm for use with *Jason*, as discussed in [2].

Environments and distributed normative objects: for certain types of environment, in particular for agent-based simulations, it would be far more convenient to use a high-level language created specifically for modelling environments for social simulation using cognitive agents [13]. This is the motivation for the ELMS language, described in [73]. That work has recently been extended [74] to allow environments to have objects containing social norms that ought to be observed only within the confines of an environment location, which is possible where an institution or organisation is situated (similarly to 'please refrain from smoking' or 'silence' signs).

Verification/model checking for AgentSpeak: a fundamental issue in computer science is whether a computer program is *correct with respect to its specification* [18]. The idea is that, when we develop a program, we have in mind some specification – the thing that we want the program to do. The program is correct if it *satisfies* its specification. The process of checking that a program satisfies its specification is called *verification*. As it turns out, verification is rather difficult for humans to do, and so a lot of effort has been devoted to getting computers to automatically verify programs. The most successful approach to computer aided verification is called *model checking* [27]. The basic idea in model checking is that a state transition graph G for a particular finite state system can be understood as a *logical model* for a *temporal logic*. If we express our specification for a system as a formula ϕ of temporal logic, then verifying that the system satisfies the specification amounts to the problem of checking that ϕ is satisfied in the model G, notationally expressed as $G \models \phi$. Model checking approaches have been developed for AgentSpeak [15, 16]. The basic idea is to *translate* the AgentSpeak program into a lower level state transition system, and then apply existing model checkers for this lower level system. The difficulty of the translation is that the interpreter for AgentSpeak must somehow be implemented in the low-level system. One target model checker for the translation was the SPIN system [53]. Using this approach, it was possible to verify properties of a range of (simple) systems. The main issue in this work is that of 'scaling up': making the techniques work on larger, realistically sized systems.

11.3 General Advice on Programming Style and Practice

By now the reader will be aware that *Jason* uses a logic-based agent programming language within a Java-based platform, and indeed makes it straightforward to use legacy Java code. In fact, it even requires the use of Java for certain advanced tasks. Because it is so easy to refer to Java code within an agent's reasoning, and there is opportunity for the use of Java code in various parts of the agent architecture and platform infrastructure, experienced Java programmers may feel tempted to abuse Java programming to avoid learning the agent programming language. This is one example of bad style in *Jason* programming that needs to be avoided. To help with this, we summarise the purpose of each usage of Java in *Jason* systems.

Environment: when you develop a multi-agent system, the notion of what is the *environment* that will be shared by autonomous agents should be very clear. If it is not, either your application is not ideal for multi-agent systems or you did not do a good job in producing your system design, which should be done with an agent-oriented software engineering methodology [52]. It often helps to start programming the agents only after there is a stable implementation of the environment model (if one is required, which is often the case, be it for testing alone). You should know precisely what the environment is like, and what are the things agents need to perceive from it so as to act appropriately, as well as what are the actions that agents will be able to perform so as to change the environment. You should only add percepts and implement actions if they are naturally part of the model of the environment as you think of it in such a way. Also, if possible, test the environment implementation before you start coding the agents to make sure there are no bugs there, as the existence of bugs in the environment makes debugging the agents much more difficult. In fact, if you have access to some other suitable language that can be used to generate Java code for simulating a multi-agent environment from a higher-level language, you could certainly combine it with *Jason* to develop the agents.

Once you have a stable implementation of the environment, you could then start implementing the agents, and this is when you run the risk of abusing the environment model to help do things you are not sure how to do in AgentSpeak. Java experts might feel great temptation to alter the environment implementation to make up 'fake' environment actions (and possibly fake percepts to send back results to the agent) so that they can do much more of the programming in Java instead of concentrating on the agents' practical reasoning. If you are a Java fanatic, be brave and resist the temptation: it will be worth it.

Internal actions: so it is not appropriate to cheat in doing extra Java programming by creating an artificial model of the environment, but surely one can use lots of internal actions to do that. Well, yes and no. If you are definitely going to cheat by implementing part of the agents' reasoning in Java, it is certainly better to do it with internal actions than by altering the environment model. Still, you should be careful not to abuse internal actions either. These should be used either for legacy code or for implementing a particular aspect of the agent's reasoning that is clearly better suited for an imperative programming style, for example certain kinds of heavy calculations, or for operations that are more efficiently done in Java *and will not remove any important declarative aspect of the agent's reasoning.*

It is important to keep in mind that almost all aspects of the agent reasoning should be done in the agent programming language. Recall that we can use prolog-like rules for theoretical reasoning (i.e. to check for consequences of what is currently believed by the agent) and plans for practical reasoning (i.e. to choose what actions to execute). Also, you may have noticed that some internal actions are used for altering some of the semantic structures used by the *Jason* interpreter. We do that in *standard* internal actions (i.e. those defined as part of *Jason*) because we know what we are doing and in the future we will make public the formal semantics of all such actions. 'Do not attempt that at home' unless you know exactly what you are doing, and even then, think twice if you really need that: our experience tells us that there is often a cleaner solution than the one that first springs to mind.

Selection functions: the AgentSpeak selection functions are effectively used for meta-level reasoning, as they can be used to do certain forms of reasoning aimed at controlling the agent's (practical) reasoning. AgentSpeak plans with higher-order variables (i.e. variables instantiated with formulæ such as goals and actions) can also be used for that purpose, to some extent. Sophisticated agents will most certainly require specific selection functions to control their reasoning (i.e. to make reasoned event/plan/intention selections). In *Jason*, we have chosen to implement the selection functions as Java methods that users can override. There are some advantages to this choice. For moderately complex applications, an agent's selection functions will be fairly simple, typically giving priority to certain events/plans, which boils down to reordering a Java list (with the structures that the Java-based interpreter in *Jason* uses). This task is very easy and efficient to do in Java, and the code for this is fairly easy to read and maintain. Further, for very complex agents, it may be necessary to use rather sophisticated techniques, for example based on decision-theory, as part of those selection functions – the work presented in [11] pointed in that direction. When an agent requires such external reasoners/schedulers to do meta-level reasoning, Java is also very useful to implement such an interface (e.g. using JNI if the external code is not in Java).

One drawback of such approach, that surely has got purists of declarative programming already horrified, is that part of the agent's reasoning has no declarative representation. To counter this problem, we have been considering, for a long time, that a language should be developed for this, which would be translated into the appropriate selection functions in Java. The language could be AgentSpeak itself, or more likely a simpler language based on AgentSpeak, and would be specific to the kinds of things that the selection functions need to consider. However, we have not yet devised such language, so this remains one area of future development for *Jason*. One final remark on selection functions is that this is one area where your design using AOSE methodologies will not have specific material to help in your development, so you will have to consider this aspect apart from the design, at least for the time being.

User customisations: one last type of Java code in software development using *Jason* is user customisations. As discussed in Chapter 7, besides changing the selection functions already covered in the previous item (and related things such as the type of belief base used), these are also used for changing the interface between the agent's overall architecture and the environment, as well as the infrastructure used for distribution and communication. In particular, changing the agent – environment interface should be used to move from a simulated environment (used to test a multi-agent system) to a real-world deployment. Keeping in mind that the interface code will be completely changed when the system is deployed should help you in deciding what you should and should not implement as part of that code. A good example of such use of customised interfaces is the 'gold miners' example that is distributed with *Jason*. We changed from that simulated environment to the 'real' CLIMA Contest competition server on the Internet by simply changing the agent architecture. This example also appears in Section 9.1, but the architecture code is not covered there due to lack of space; the complete code can be found in the webpage for this book at URL `http://jason.sf.net/jBook`. Finally, *Jason* is built in such a way that the infrastructure used for agent distribution over a computer network and for agent communication can also be changed in a straightforward way. We have recently finished integrating the JADE [9] infrastructure into *Jason*, so that we can have *Jason* applications running on top of JADE, which is quickly becoming a *de facto* standard for agent distribution and communication.

As a final word, recall that this book has a web page, available at `http://jason.sf.net/jBook`, where we expect to make available a variety of resources to accompany this book. *Jason* itself can be downloaded from http://jason.sf.net; you can also have access to the user-support email lists from there.

Appendix: Reference Guide

A.1 EBNF for the Agent Language

```
agent            →  init_bels init_goals plans
init_bels        →  beliefs rules
beliefs          →  ( literal "." )*
rules            →  ( literal ":-" log_expr ".")*
init_goals       →  ( "!" literal "." )*
plans            →  ( plan )*
plan             →  [ "@" atomic_formula ] triggering_event
                    [ ":" context ]
                    [ "<-" body ] "."
triggering_event →  ( "+" | "-" ) [ "!" | "?" ] literal
literal          →  [ "~" ] atomic_formula
context          →  log_expr  |  "true"
log_expr         →  simple_log_expr
                 |  "not" log_expr
                 |  log_expr "&" log_expr
                 |  log_expr "|" log_expr
                 |  "(" log_expr ")"
simple_log_expr  →  ( literal | rel_expr | <VAR> )
body             →  body_formula ( ";" body_formula )*
                 |  "true"
body_formula     →  ( "!" | "!!" | "?" | "+" | "-" | "-+" ) literal
                 |  atomic_formula
                 |  <VAR>
                 |  rel_expr
atomic_formula   →  ( <ATOM> | <VAR> )
                    [ "(" list_of_terms ")" ]
                    [ "[" list_of_terms "]" ]
list_of_terms    →  term ( "," term )*
term             →  literal
                 |  list
                 |  arithm_expr
                 |  <VAR>
                 |  <STRING>
```

```
list          →  "[" [ term ( "," term )* [ "|" ( list | <VAR> ) ] ] "]"

rel_expr      →  rel_term
                 [ ("<"|"<="|">"|">="|"=="|"\=="|"="|"=..") rel_term ]+
rel_term      →  (literal|arithm_expr)
arithm_expr   →  arithm_term
                 [ ( "+" | "-" | "*" | "**" | "/" | "div" | "mod" )
                   arithm_term ]*
arithm_term   →  <NUMBER>
              |  <VAR>
              |  "-" arithm_term
              |  "(" arithm_expr ")"
```

It is worth noting that this grammar has been slightly simplified to make it more readable; the actual grammar used to generate the parser used by *Jason* is available in the distribution (file doc/AS2JavaParser.html). On particular thing that this grammar does not reflect is that initial beliefs, initial goals and plans do not necessarily need to appear in this order, *even though it is very much recommended that this order is kept*, at least within an AgentSpeak file. One situation, for example, where these constructs will unavoidably appear interleaved in the final AgentSpeak code for an agent is when various AgentSpeak files are included in an agent program and each contributes to the initial beliefs, initial goals and plan library.

A.2 EBNF for the Multi-Agent Systems Language

```
mas              →  "MAS" <ID> "{"
                       [ infrastructure ]
                       [ environment ]
                       [ exec_control ]
                       agents
                    "}"
infrastructure   →  "infrastructure" ":" <ID>
environment      →  "environment" ":" <ID> [ "at" <ID> ]
exec_control     →  "executionControl" ":" <ID> [ "at" <ID> ]
agents           →  "agents" ":" ( agent )+
agent            →  <ASID>
                    [ filename ]
                    [ options ]
                    [ "agentArchClass" <ID> ]
                    [ "beliefBaseClass" <ID> ]
                    [ "agentClass" <ID> ]
                    [ "#" <NUMBER> ]
                    [ "at" <ID> ]
                    ";"
filename         →  [ <PATH> ] <ID>
options          →  "[" option ( "," option )* "]"
option           →  "events" "=" ( "discard" | "requeue" | "retrieve" )
                 |  "intBels" "=" ( "sameFocus" | "newFocus" )
                 |  "nrcbp" "=" <NUMBER>
```

```
|   "verbose" "=" <NUMBER>
|   <ID> "=" ( <ID> | <STRING> | <NUMBER> )
```

The grammar used to generate the parser used by *Jason* is available in the distribution (file `doc/MAS2JavaParser.html`).

A.3   Standard Internal Actions

This section describes all standard internal actions available in the *Jason* distribution, their arguments, and examples. They are formatted as shown below and organised by purpose (communication, plan library management, etc.).

.internal action name	<description>.
	1. <modifier> **<first argument name>** (<type of the argument>): <description>.
	Arguments in *italic* are optional.
	If the modifier is '+', the argument cannot be a free variable, it is a value being passed to the internal action.
	If the modifier is '-', the argument should be a variable (i.e., it should not be a ground term), the return value is likely to be unified with this variable.
	If the modifier is '+/-', the argument can be used both as for '+' and for '-'.
	2. second argument.
	3. ...

Examples:

- <list of examples>.

Communication

.broadcast	broadcasts a message to all known agents.
	1. + **ilf** (atom): the illocutionary force of the message (tell, achieve, ...).
	2. + **message** (literal): the content of the message.

Examples:

- `.broadcast(tell,value(10))`: sends `value(10)` as a 'tell' message to all known agents in the society.

.my_name   gets the agent's unique identification in the multi-agent system. This identification is given by the runtime infrastructure of the system (centralised, saci, jade, ...).

1. +/- **name** (atom): if this is a variable, unifies the agent name and the variable; if it is an atom, succeeds if the atom is equal to the agent's name.

Examples:

- `.my_name(N)`: unifies N with the agent's name.

.send   sends a message to an agent.
Messages with an `ask` illocutionary force can optionally have arguments answer and timeout. If they are given, `.send` suspends the intention until an answer is received and unified with the answer argument, or the message request times out as specified by timeout. Otherwise, the intention is not suspended and the answer (which is a tell message) produces a belief addition event as usual.

1. + **receiver** (atom, string, or list): the receiver of the message. It is the unique name of the agent that will receive the message (or list of names).

2. + **ilf** (atom): the illocutionary force of the message (tell, achieve, ...).

3. + **message** (literal): the content of the message.

4. + *answer* (any term): the answer of an ask message (for performatives AskOne and AskAll).

5. + *timeout* (number): timeout (in milliseconds) when waiting for an ask answer.

Examples (suppose that agent `jomi` is sending the messages):

- `.send(rafael,tell,value(10))`: sends `value(10)` to the agent named `rafael`. The literal `value(10)[source(jomi)]` will be added as a belief in `rafael`'s belief base.
- `.send(rafael,achieve,go(10,30)`: sends `go(10,30)` to the agent named `rafael`. When `rafael` receives this message, an event `<+!go(10,30)[source(jomi)],T>` will be added in `rafael`'s event queue.
- `.send(rafael,askOne,value(beer,X))`: sends `value(beer,X)` to the agent named `rafael`. This `.send` does not suspend `jomi`'s intention. An event `+value(beer,10)[source(rafael)]` is generated in `jomi`'s side when `rafael` answers the ask.
- `.send(rafael,askOne,value(beer,X),A)`: sends `value(beer,X)` to the agent named `rafael`. This action suspends `jomi`'s intention until `rafael`'s answer is received. The answer (something like `value(beer,10)`) unifies with `A`.
- `.send(rafael,askOne,value(beer,X),A,2000)`: as in the previous example, but agent `jomi` waits for 2 seconds. If no message is received by then, `A` unifies with `timeout`.

List and String Manipulation

.concat	concatenates strings or lists.

1. **+ arg_0 ... arg_{n-1}** (any term): the terms to be concatenated. Parameters that are not strings are concatenated using the toString method of their class.

2. **+/- arg_n** (string or list): the result of the concatenation.

Examples:

- `.concat("a","b",X)`: `X` unifies with 'ab'.
- `.concat("a","b","a")`: false.
- `.concat("a b",1,a,X)`: `X` unifies with 'a b1a'.
- `.concat("a", "b", "c", "d", X)`: `X` unifies with 'abcd'.
- `.concat([a,b,c],[d,e],[f,g],X)`: `X` unifies with `[a,b,c,d,e,f,g]`.

.length   gets the length of strings or lists.

      1. **+ argument** (string or list): the term whose length is to be determined.

      2. **+/- length** (number).

Examples:

- `.length("abc",X)`: X unifies with 3.
- `.length([a,b],X)`: X unifies with 2.
- `.length("a",2)`: false.

.max   gets the maximum value within a list of terms, using the 'natural' order for each type of term.

      1. **+ terms** (list): the list where to find the maximum term.

      2. **+/- maximum** (term).

Examples:

- `.max([c,a,b],X)`: unifies X with c.
- `.max([b,c,10,g,f(10),5,f(4)],X)`: unifies X with f(10).
- `.max([3,2,5],2)`: false.
- `.max([3,2,5],5)`: true.
- `.max([],X)`: false.

.member   checks if some term is in a list. If the term is a free variable, this internal action backtracks all possible values for the list.

      1. **+/- member** (term): the term to be checked.

      2. **+ list** (list): the list where the term should be in.

Examples:

- `.member(c,[a,b,c])`: true.
- `.member(3,[a,b,c])`: false.
- `.member(X,[a,b,c])`: unifies X with any member of the list.

.min   gets the minimum value of a list of terms, using the
                            'natural' order for each type of term.

    1. + **terms** (list): the list where to find the mini-
                            mum term.

    2. +/- **minimum** (term).

Examples:

- `.min([c,a,b],X)`: unifies X with a.
- `.min([b,c,10,g,f(10),5,f(4)],X)`: unifies X with 5.
- `.min([3,2,5],2)`: true.
- `.min([3,2,5],5)`: false.
- `.min([],X)`: false.

.nth   gets the *n*th term of a list.

    1. + **index** (integer): the position of the term, the
                            first term is at position 0.

    2. + **list** (list): the list where to get the term from.

    3. +/- **term** (term): the term at position *index* in
                            the *list*.

Examples:

- `.nth(0,[a,b,c],X)`: unifies X with a.
- `.nth(2,[a,b,c],X)`: unifies X with c.
- `.nth(0,[a,b,c],d)`: false.
- `.nth(0,[a,b,c],a)`: true.
- `.nth(5,[a,b,c],X)`: error.

.sort   sorts a list of terms. The 'natural' order for each type
                            of terms is used. Between different types of terms,
                            the following order is used: numbers < atoms <
                            structures < lists.

    1. + **unordered list** (list): the list to be sorted.

    2. +/- **ordered list** (list): the sorted list.

Examples:

- `.sort([c,a,b],X)`: X unifies with `[a,b,c]`.
- `.sort([b,c,10,g,casa,f(10),[3,4],5,[3,10],f(4)],X)`: X unifies with `[5,10,b,c,casa,g,f(4),f(10),[3,4],[3,10]]`.
- `.sort([3,2,5],[2,3,5])`: true.
- `.sort([3,2,5],[a,b,c])`: false.

.substring

checks if a string is sub-string of another string. The arguments can be other kinds of term, in which case the toString() of the term is used. If 'position' is a free variable, the internal action backtracks all possible values for the positions where the sub-string occurs in the string.

1. + **sub-string** (any term).

2. + **string** (any term).

3. +/- *position* (integer): the position of the string where the sub-string occurs.

Examples:

- `.substring("b","aaa")`: false.
- `.substring("b","aaa",X)`: false.
- `.substring("a","bbacc")`: true.
- `.substring("a","abbacca",X)`: true and X unifies with 0, 3, and 6.
- `.substring("a","bbacc",0)`: false.
- `.substring(a(10),b(t1,a(10)),X)`: true and X unifies with 5.
- `.substring(a(10),b("t1,a(10)"),X)`: true and X unifies with 6.

Plan Library Manipulation

.add_plan

adds plan(s) to the agent's plan library.

1. + **plans** (string or list): the string representing the plan to be added. If it is a list, each string in the list will be parsed into an AgentSpeak plan and added to the plan library. The syntax of the code within the string is the same as ordinary AgentSpeak code.

2. + *source* (structure): the source of the plan. The default value for the source is `self`.

Examples:

- `.add_plan("+b : true <- .print(b).")`: adds the plan `+b : true <- .print(b).` to the agent's plan library with a plan label annotated with `source(self)`.
- `.add_plan("+b : true <- .print(b).", rafa)`: same as the previous example, but the source of the plan is agent `rafa`.
- `.add_plan(["+b : true <- .print(b).", "+b : bel <- .print(bel)."], rafa)`: adds both plans with `rafa` as their sources.

.plan_label   unifies the first argument with a string representing the plan labelled with the second argument within the agent's plan library.

     1. - **plan** (string): the string representing the plan.

     2. + **label** (structure): the label of that plan.

Examples:

- `.plan_label(P,p1)`: unifies `P` with the string representation of the plan labelled `p1`.

.relevant_plans   gets all relevant plans for some triggering event. This internal action is used, for example, to answer 'AskHow' messages.

     1. + **trigger** (string): the string representing the triggering event.

     2. - **plans** (list): the list of strings with the code of the relevant plans.

Examples:

- `.relevant_plans("+!go(X,Y)",L)`: unifies `L` with a list of all plans in the agent's plan library that are relevant for the triggering event `+!go(X,Y)`.

.remove_plan   removes plans from the agent's plan library.

1. **+ label(s)** (structure or list of structures): the label of the plan to be removed. If this parameter is a list of labels, all plans of this list are removed.

2. **+ *source*** (atom): the source of the plan to be removed. The default value is `self`.

Examples:

- `.remove_plan(l1)`: removes the plan identified by label `l1[source(self)]`.
- `.remove_plan(l1,bob)`: removes the plan identified by label `l1[source(bob)]`. Note that a plan with a source like that was probably added to the plan library by a TellHow message.
- `.remove_plan([l1,l2,l3])`: removes the plans identified by labels `l1[source(self)]`, `l2[source(self)]`, and `l3[source(self)]`.
- `.remove_plan([l1,l2,l3],bob)`: removes the plans identified by labels `l1[source(bob)]`, `l2[source(bob)]`, and `l3[source(bob)]`.

BDI

.current_intention   returns a description of the current intention. It is useful for plans that need to inspect the current intention.

The description of the intention has the following form:
```
intention(<id>,<stack of intended means>)
```

where each intended means has the form:
```
im(<plan label>,<list of body literals>)
```

For example:
```
intention(1,
  [
    im("l__6",[".current_intention(I)"]),
    im("l__5",[".fail"]),
    im("l__4",["!g5(X)",".print(endg4)"]),
    ...
  ]).
```

Examples:

- `.current_intention(X)`: X unifies with the descriptions of the current intention (i.e. the intention that executed this internal action).

.desire   checks whether the argument is a desire: it is a desire either if there is an event with `+!<desire>` as triggering event or it is a goal in one of the agent's intentions.

Examples:

- `.desire(go(1,3))`: true if `go(1,3)` is a desire of the agent.

.drop_all_desires   removes all desires of the agent, including those already committed to (i.e. the intentions). No event is produced.

This action changes the agent's circumstance structure by simply emptying the whole set of events (E) then doing as described in `drop_all_intentions`.

Examples:

- `.drop_all_desires`.

.drop_all_events   removes all desires that the agent has not yet committed to. No event is produced.

This action changes the agent's circumstance structure by simply emptying the whole set of events (E). This action is complementary to `.drop_all_desires` and `.drop_all_intentions`, in case all entries are to be removed from the set of events but *not* from the set of intentions.

Examples:

- `.drop_all_events`.

.drop_all_intentions	removes all intentions from the agent's set of intentions (even suspended intentions are removed). No event is produced.

This action changes the agent's circumstance structure by simply emptying the whole set of intentions (I), pending actions (PA), pending intentions (PI), and events in E that are not external events (thus generated by intentions).

Examples:

- `.drop_all_intentions.`

.drop_desire	removes a desire D from the agent circumstance. This internal action simply removes all `+!D` entries (those for which `.desire(D)` would succeed) from both the set of events and the set of intentions. No event is produced as a consequence of dropping desires.

1. + **desire** (literal): the literal to be unified with the desires that are to be dropped.

Examples:

- `.drop_desire(go(X,3))`: removes desires such as `<+!go(1,3),_>` from the set of events and intentions having plans with triggering events such as `+!go(1,3)`.

.drop_event	removes events D from the agent circumstance. This internal action simply removes all `+!D` entries (those for which `.desire(D)` would succeed) *from the set of events only*; this action is complementary to `.drop_desire` and `.drop_intention`, if a goal is to be removed only from the set of events and *not* from the set of intentions. No event is produced as a consequence of dropping desires from the set of events.

1. + **desire** (literal): the literal to be unified with the desires that are to be dropped.

Examples:

- `.drop_event(go(X,Y))`: removes events such as `<+!go(1,3),_>` from the set of events.

.drop_intention	removes intentions `I` from the set of intentions of the agent (suspended intentions are also considered). No event is produced.

> 1. **+ intention** (literal): the literal to be unified with the intentions that are to be dropped.

Examples:

- `.drop_intention(go(1,3))`: removes an intention having a plan with triggering event `+!g(1,3)` in the agent's current circumstance.

.fail_goal	aborts goals `G` in the agent circumstance as if a plan for such a goal had failed. Assuming that one of the plans requiring `G` was `G0 <- !G; ...`, an event `-!G0` is generated. If `G` was triggered by `!!G` (and therefore not a subgoal, as happens also when an 'achieve' performative is used), the generated event is `-!G`. A literal `G` is a goal if there is a triggering event `+!G` in any plan within any intention; also note that intentions can be suspended, hence appearing in sets *E*, *PA*, or *PI* of the agent's circumstance as well (this internal action is explained in detail in Section 8.1).

> 1. **+ goal** (literal).

Examples:

- `.fail_goal(go(X,3))`: aborts any attempt to achieve goals such as `!go(1,3)` as if a plan for it had failed. Assuming that it is a subgoal in the plan `get_gold(X,Y) <- go(X,Y); pick.`, the generated event is `-!get_gold(1,3)`.

.intend	checks whether the argument is an intention: it is an intention if there is a triggering event `+!<intention>` in any plan within an intention; just note that intentions can be suspended and appear in E, PA, and PI as well.

> 1. **+ intention** (literal).

Examples:

- .intend(go(1,3)): is true if a plan with triggering event +!go(1,3) appears in an intention of the agent.

.succeed_goal   removes goals G from the agent circumstance as if the plan for such goal had successfully finished. G is a goal if there is a triggering event +!G in any plan within any intention; also note that intentions can be suspended hence appearing in E, PA, or PI as well (this internal action is explained in detail in Section 8.1).

1. + **goal** (literal).

Examples:

- .succeed_goal(go(X,3)): aborts any attempt to achieve goals such as !go(1,3) as if they had already been achieved.

Term Type Identification

.atom   checks whether the argument is an atom (a structure with arity 0), for example p. Numbers, strings and free variables are *not* atoms.

Examples:

- .atom(b(10)): false.
- .atom(b): true.
- .atom(~b): false.
- .atom(10): false.
- .atom("home page"): false.
- .atom(X): only true if X is bound to an atom.
- .atom(a(X)): false.
- .atom(a[X]): false.
- .atom([a,b,c]): false.
- .atom([a,b,c(X)]): false.

.ground   checks whether the argument is ground, i.e. it has no free variables. Numbers, strings and atoms are always ground.

Examples:

- `.ground(b(10))`: true.
- `.ground(10)`: true.
- `.ground(X)`: false if X is free or bound to a term with free variables.
- `.ground(a(X))`: false if X is free or bound to a term with free variables.
- `.ground([a,b,c])`: true.
- `.ground([a,b,c(X)])`: false if X is free or bound to a term with free variables.

.literal   checks whether the argument is a literal, e.g. `p`, `p(1)`, `p(1)[a,b]`, `~p(1)[a,b]`.

Examples:

- `.literal(b(10))`: true.
- `.literal(b)`: true.
- `.literal(10)`: false.
- `.literal("Jason")`: false.
- `.literal(X)`: false if X is free, true if X is bound to a literal.
- `.literal(a(X))`: true.
- `.literal([a,b,c])`: false.
- `.literal([a,b,c(X)])`: false.

.list   checks whether the argument is a list.

Examples:

- `.list([a,b,c])`: true.
- `.list([a,b,c(X)])`: true.
- `.list(b(10))`: false.
- `.list(10)`: false.
- `.list("home page")`: false.
- `.list(X)`: false if X is free, true if X is bound to a list.
- `.list(a(X))`: false.

.number   checks whether the argument is a number.

Examples:

- `.number(10)`: true.
- `.number(10.34)`: true.
- `.number(b(10))`: false.
- `.number("home page")`: false.
- `.number(X)`: false if X is free, true if X is bound to a number.

.string   checks whether the argument is a string.

Examples:

- `.string("home page")`: true.
- `.string(b(10))`: false.
- `.string(b)`: false.
- `.string(X)`: false if X is free, true if X is bound to a string.

.structure   checks whether the argument is a structure.

Examples:

- `.structure(b(10))`: true.
- `.structure(b)`: true.
- `.structure(10)`: false.
- `.structure("home page")`: false.
- `.structure(X)`: false if X is free, true if X is bound to a structure.
- `.structure(a(X))`: true.
- `.structure([a,b,c])`: true.
- `.structure([a,b,c(X)])`: true.

Miscellaneous

.abolish   removes all beliefs that match the argument. As for the '−' operator, an event will be generated for each deletion.

1. **+ belief** (literal): the 'pattern' for what should be removed.

Examples:

- `.abolish(b(_))`: remove all `b/1` beliefs, regardless of the argument value.
- `.abolish(c(_,2))`: remove all `c/2` beliefs where the second argument is 2.
- `.abolish(c(_,_)[source(ag1)])`: remove all `c/2` beliefs that have `ag1` as one of the sources.

.add_annot   adds an annotation to a literal.

1. **+ belief(s)** (literal or list): the literal where the annotation is to be added. If this parameter is a list, all literals in the list will have the annotation added.

2. **+ annotation** (structure).

3. **+/- annotated beliefs(s)** (literal or list): this argument unifies with the result of the annotation addition.

Examples:

- `.add_annot(a,source(jomi),B)`: B unifies with `a[source(jomi)]`.
- `.add_annot(a,source(jomi),b[jomi])`: fails because the result of the addition does not unify with the third argument.
- `.add_annot([a1,a2], source(jomi), B)`: B unifies with `[a1[source(jomi)], a2[source(jomi)]]`.

.at   creates an event at some time in the future. This command is based on the unix 'at' command, although not fully implemented yet.

1. **+ when** (string): the time for the event to be generated.

   The syntax of this string in the current implementation has the following format:
   ```
   now + <number> [<time_unit>]
   ```
   where `<time_unit>` can be 's' or 'second(s)', 'm' or 'minute(s)', 'h' or 'hour(s)', 'd' or 'day(s)'. The default is milliseconds.

2. **+ event** (string): the event to be created. The string will be parsed as a triggering event.

Examples:

- `.at("now +3 minutes", "+!g")`: generates the event `+!g` 3 minutes from now.
- `.at("now +1 m", "+!g")`
- `.at("now +2 h", "+!g")`

.count   counts the number of occurrences of a particular belief (pattern) in the agent's belief base.

       1. **+ belief** (literal): the belief to be counted.

       2. **+/- quantity** (number): the number of occurrences of the belief.

Examples:

- `.count(a(2,_),N)`: counts the number of beliefs that unify with `a(2,_)`; `N` unifies with this quantity.
- `.count(a(2,_),5)`: succeeds if the BB has exactly 5 beliefs that unify with `a(2,_)`.

.create_agent   creates another agent using the referred AgentSpeak source code (see also `kill_agent` on page 254).

       1. **+ name** (atom): the name for the new agent.

       2. **+ source** (string): path to the file where the AgentSpeak code for the new agent can be found.

       3. **+ *customisations*** (list): list of optional parameters as agent class, architecture and belief base.

Examples:

- `.create_agent(bob,"/tmp/x.asl")`: creates an agent named 'bob' from the source file in `/tmp/x.asl`.
- `.create_agent(bob,"x.asl", [agentClass(myp.MyAgent)])`: creates the agent with customised agent class `myp.MyAgent`.
- `.create_agent(bob,"x.asl", [agentArchClass(myp.MyArch)])`: creates the agent with customised architecture class `myp.MyArch`.
- `.create_agent(bob, "x.asl", [beliefBaseClass(jason.bb.TextPersistentBB)])`: creates the agent with customised belief base `jason.bb.TextPersistentBB`.

- `.create_agent(bob,"x.asl", [agentClass(myp.MyAgent),`
`agentArchClass(myp.MyArch),beliefBaseClass(jason.bb.`
`TextPersistentBB)])`: creates the agent with agent, architecture and belief base customised.

.date   gets the current date (year, month, and day of the month).

     1. +/- **year** (number): the year.

     2. +/- **month** (number): the month (1–12).

     3. +/- **day** (number): the day (1–31).

Examples:

- `.date(Y,M,D)`: unifies Y with the current year, M with the current month, and D with the current day.
- `.date(2006,12,30)`: succeeds if the action is run on 30/Dec/2006 and fails otherwise.

.fail   fails the intention where it is run (an internal action that always returns false).

.findall   builds a list of all instantiations of some `term` which make some `query` a logical consequence of the agent's BB. Unlike in Prolog, the second argument cannot be a conjunction.

     1. + **term** (variable or structure): the variable or structure whose instances will 'populate' the list.

     2. + **query** (literal): the literal to match against the belief base.

     3. +/- **result** (list): the resulting populated list.

Examples (assuming the BB is currently {a(30),a(20),b(1,2),b(3,4),b(5,6)}):

- `.findall(X,a(X),L)`: L unifies with `[30,20]`.
- `.findall(c(Y,X),b(X,Y),L)`: L unifies with `[c(2,1),c(4,3), c(6,5)]`.

.kill_agent	kills the agent whose name is given as parameter. This is a provisional internal action to be used while more adequate mechanisms for creating and killing agents are being developed. In particular, note that an agent can kill any other agent, without any consideration of permissions, etc.! It is the programmer's responsibility to use this action (see also `create_agent` on page 252).

1. + **name** (atom): the name of the agent to be killed.

Examples:

- `.kill_agent(bob)`: kills the agent named 'bob'.

.perceive	forces the agent architecture to do perception of the environment immediately. It is normally used when the number of reasoning cycles before perception takes place was changed (this is normally at every cycle) .

.random	unifies the argument with a random number between 0 and 1.

1. - **value** (number): the variable to receive the random value.

.stopMAS	aborts the execution of all agents in the multi-agent system (and any simulated environment too).

.time	gets the current time (hours, minutes, and seconds).

1. +/- **hours** (number): the hours (0–23).

2. +/- **minutes** (number): the minutes (0–59).

3. +/- **seconds** (number): the seconds (0–59).

Examples:

- `.time(H,M,S)`: unifies H with the current hour, M with the current minutes, and S with the current seconds.
- `.time(15,_,_)`: succeeds if it is now 3 p.m. or a little later but not yet 4 p.m.

.wait	suspends the intention for the time specified or until some event happens.

          1. + *event* (string): the event to wait for. The events are strings in AgentSpeak syntax, e.g. `+bel(33)`, `+!go(X,Y)`

          2. + *timeout* (number): timeout in mileseconds.

Examples:

- `.wait(1000)`: suspend the intention for 1 second.
- `.wait("+b(1)")`: suspend the intention until the belief `b(1)` is added in the belief base.
- `.wait("+!g", 2000)`: suspend the intention until the goal `g` is triggered or 2 seconds have passed, whichever happens first.

A.4 Pre-Defined Annotations

For beliefs, the `source` annotation is the only one that has a special meaning for *Jason*; see Section 3.1 for more details. The available pre-defined plan annotations are as follows.

atomic: if an instance of a plan with an `atomic` annotation is chosen for execution by the intention selection function, this intention will be selected for execution in the subsequent reasoning cycles *until that plan is finished* — note that this overrides any decisions of the intention selection function in subsequent cycles (in fact, the interpreter does not even call the intention selection function after an `atomic` plan began to be executed). This is quite handy in cases where the programmer needs to guarantee that no other intention of that agent will be executed in between the execution of the formulæ in the body of that plan.

breakpoint: this is very handy in debugging — if the `debug` mode is being used (see Section 4.3), as soon as any of the agents start executing an intention with an instance of a plan that has a `breakpoint` annotation, the execution stops and the control goes back to the user, who can then run the system step-by-step or resume normal execution.

`all_unifs:` this is used to include all possible unifications that make a plan applicable in the set of applicable plans. Normally, for one given plan, only the first unification found is included in the set of applicable plans. In normal circumstances, the applicable-plan selection function is used to choose between *different* plans, all of which are applicable. If you have created a plan for which you want the applicable-plan selection function to consider which is the best variable substitution for the plan to be used as intended means for the given event, then you should include this special annotation in the plan label.

`priority:` is a term (of arity 1) which can be used by the default plan selection function to give a simple priority order for applicable plans. The higher the integer number given as parameters, the higher the plan's priority in the case of multiple applicable plans. Recall that you can implement more sophisticated selection functions and still make use of this (and other user-defined) annotation if you want. (Note that this is as yet unavailable, but it is a reserved plan annotation for this purpose.)

A.5 Pre-Processing Directives

The pre-processing directives currently available are as follows. Recall that users can define their own, as explained in Section 7.5.

`{include("⟨file_name⟩")}`

> The parameter to this directive is a string with the file name (and path if necessary) for an AgentSpeak source code to be included in the current AgentSpeak file.

Differently from 'include', all other directives, which are for plan patterns, have a `{begin ...}` ... `{end}` structure.

`{begin dg(⟨goal⟩)}`
>   `⟨plans⟩`
`{end}`

> This is the 'declarative goal' pattern described in Section 8.2.

`{begin bdg(⟨goal⟩)}`
>   `⟨plans⟩`
`{end}`

This is the 'backtracking declarative goal' pattern described in Section 8.2. Recall that further patterns can be used in the body of the directive where the ⟨*plans*⟩ are given.

```
{begin ebdg(⟨goal⟩))}
    ⟨plans⟩
{end}
```

This is the 'exclusive backtracking declarative goal' pattern described in Section 8.2.

```
{begin bc(⟨goal⟩,⟨F⟩))}
    ⟨plans⟩
{end}
```

This is used to add a 'blind commitment' strategy to a given goal. *F* is the name of another pattern to be used for the goal itself, the one for which the commitment strategy will be added; it is typically BDG, but could be any other goal type.

```
{begin smc(⟨goal⟩,⟨f⟩))}
    ⟨plans⟩
{end}
```

This is the pattern to add a 'single-minded commitment' strategy to a given goal; *f* is the *failure condition*.

```
{begin rc(⟨goal⟩,⟨m⟩))}
    ⟨plans⟩
{end}
```

This is the pattern to add a 'relativised commitment' strategy to a given goal; *m* is the *motivation* for the goal.

```
{begin omc(⟨goal⟩,⟨f⟩,⟨m⟩))}
    ⟨plans⟩
{end}
```

This is the pattern to add an 'open-minded commitment' strategy to a given goal; *f* and *m* are as in the previous two patterns.

```
{begin mg(⟨goal⟩,⟨F⟩))}
    ⟨plans⟩
{end}
```

This is the 'maintenance goal' pattern; F is the name of another pattern to be used for the achievement goal, if the maintenance goal fails. Recall that this is a reactive form of maintenance goal, rather than pro-active. It causes the agent to act to return to a desired state after things have already gone wrong.

```
{begin sga(⟨t⟩,⟨c⟩,⟨goal⟩))}
    ⟨plans⟩
{end}
```

This is the 'sequenced goal adoption' pattern; the pattern prevents multiple instances of the same plan to be simultaneously adopted, where t is the triggering event and c is the context of the plan.

A.6   Interpreter Configuration

The available interpreter configurations that can be used for each individual agent in the .mas2j file are as follows. The underlined option is the one used by default.

events: options are <u>discard</u>, requeue or retrieve; the discard option means that external events for which there are no applicable plans are discarded (a warning is printed out to the console), whereas the requeue option is used when such events should be inserted back at the end of the list of events that the agent needs to handle. An option retrieve is also available; when this option is selected, the user-defined selectOption function is called even if the set of relevant/applicable plans is empty. This can be used, for example, for allowing agents to request plans from other agents who may have the necessary know-how that the agent currently lacks, as proposed in [6].

intBels: options are either <u>sameFocus</u> or newFocus; when internal beliefs are added or removed explicitly within the body of a plan, if the associated event is a triggering event for a plan, the intended means resulting from the applicable plan chosen for that event can be pushed on top of the intention (i.e. focus of attention) which generated the event, or it could be treated as an external event (as the addition or deletions of belief from perception of the environment), creating a new focus of attention. Because this was not considered in the original version of the language, and it seems to us that both options can be useful, depending on the application, we left this as an option for the user.

nrcbp: number of reasoning cycles before perception. Not all reasoning cycles will lead the agent to execute an action that changes the environment, and unless the environment is extremely dynamic, the environment might not have changed at all while the agent has done a single reasoning cycle. Besides, for realistic applications, perceiving the environment and obtaining a symbolic representation of it can be a very computationally expensive task. Therefore, in many cases it might be interesting to configure the number of reasoning cycles the agent will have before the perceive method is actually called to get updated sensorial information. This can be done easily by setting the nrcbp interpreter configuration option. The default is 1, as in the original (abstract) conception of AgentSpeak(L) – that is, the environment is perceived at every single reasoning cycle.

The parameter could, if required, be given an artificially high number so as to prevent perception and belief revision ever happening 'spontaneously'. If this is done, programmers need to code their plans to actively perceive the environment and do belief revision, as happens in various other agent-oriented programming languages. See page 254, where the internal action .perceive() is described.

verbose: a number between 0 and 2 should be specified. The higher the number, the more information about that agent (or agents if the number of instances is greater than 1) is printed out in the *Jason* console. The default is in fact 1, not 0; verbose 1 prints out only the actions that agents perform in the environment and the messages exchanged between them. Verbose 0 prints out only messages from the .print and .println internal actions. Verbose 2 is the 'debugging' mode, so the messages are very detailed.

user settings: Users can create their own settings in the agent declaration, for example:

```
... agents: ag1 [verbose=2,file="an.xml",value=45];
```

These extra parameters are stored in the Settings class and can be checked within the programmer's classes with the getUserParameter method, for example:

```
getSettings().getUserParameter("file");
```

This type of configuration is typically only useful for advanced users.

References

[1] Alberti M, Chesani F, Gavanelli M, Lamma E, Mello P and Torroni P 2005 The SOCS computational logic approach to the specification and verification of agent societies. In *Global Computing, IST/FET International Workshop, GC 2004, Rovereto, 9 – 12 March 2004, Revised Selected Papers* (ed. Priami C and Quaglia P), vol. 3267 of *LNCS*, pp. 314 – 339. Springer, Berlin.

[2] Alechina N, Bordini RH, Hübner JF, Jago M and Logan B 2007 Automating belief revision for AgentSpeak. In *Proceedings of the Fourth International Workshop on Declarative Agent Languages and Technologies (DALT 2006), held with AAMAS 2006, 8 May, Hakodate* (ed. Baldoni M and Endriss U), vol. 4327 of *Lecture Notes in Computer Science*, pp. 61 – 77. Springer, Berlin, series.

[3] Alechina N, Jago M and Logan B 2006 Resource-bounded belief revision and contraction. In *Declarative Agent Languages and Technologies III, Third International Workshop, DALT 2005, Utrecht, 25 July 2005, Selected and Revised Papers* (ed. Baldoni M, Endriss U, Omicini A and Torroni P), vol. 3904 of *LNAI*, pp. 141 – 154. Springer, Berlin.

[4] Allen JF, Hendler J and Tate A (eds) 1990 *Readings in Planning*. Morgan Kaufmann, San Mateo, CA.

[5] Ancona D and Mascardi V 2004 Coo-BDI: Extending the BDI model with cooperativity. In *Declarative Agent Languages and Technologies, First International Workshop, DALT 2003, Melbourne, 15 July 2003, Revised Selected and Invited Papers* (ed. Leite JA, Omicini A, Sterling L and Torroni P), vol. 2990 of *LNAI*, pp. 109 – 134. Springer.

[6] Ancona D, Mascardi V, Hübner JF and Bordini RH 2004 Coo-AgentSpeak: cooperation in AgentSpeak through plan exchange. In *Proceedings of the Third International Joint Conference on Autonomous Agents and Multi-Agent Systems (AAMAS-2004), New York, 19 – 23 July* (ed. Jennings NR, Sierra C, Sonenberg L and Tambe M), pp. 698 – 705. ACM Press, New York.

[7] Antoniou G and Harmelen F 2004 *A Semantic Web Primer*. The MIT Press, Cambridge, MA.

[8] Austin JL 1962 *How to Do Things With Words*. Oxford University Press, Oxford.

[9] Bellifemine F, Bergenti F, Caire G and Poggi A 2005 JADE – a Java agent development framework. In [14], chapter 5, pp. 125 – 147.

[10] Bordini RH and Moreira ÁF 2004 Proving BDI properties of agent-oriented programming languages: the asymmetry thesis principles in AgentSpeak(L). *Annals of
Mathematics and Artificial Intelligence* **42**(1 – 3), 197 – 226. Special Issue on Computational Logic in Multi-Agent Systems.

[11] Bordini RH, Bazzan ALC, Jannone RO, Basso DM, Vicari RM and Lesser VR
2002 AgentSpeak(XL): efficient intention selection in BDI agents via decision-
theoretic task scheduling. In *Proceedings of the First International Joint Conference
on Autonomous Agents and Multi-Agent Systems (AAMAS-2002), 15 – 19 July, Bologna*
(ed. Castelfranchi C and Johnson WL), pp. 1294 – 1302. ACM Press, New York.

[12] Bordini RH, Braubach L, Dastani M, Seghrouchni AEF, Gomez-Sanz JJ, Leite J,
O'Hare G, Pokahr A and Ricci A 2006a A survey of programming languages and
platforms for multi-agent systems. *Informatica* **30**(1), 33 – 44. Available online at
http://ai.ijs.si/informatica/vols/vol30.html#No1.

[13] Bordini RH, da Rocha Costa AC, Hübner JF, Moreira ÁF, Okuyama FY and
Vieira R 2005a MAS-SOC: a social simulation platform based on agent-oriented
programming. *Journal of Artificial Societies and Social Simulation*. JASSS Forum,
<http://jasss.soc.surrey.ac.uk/8/3/7.html>.

[14] Bordini RH, Dastani M, Dix J and El Fallah Seghrouchni A (eds) 2005b *Multi-Agent
Programming: Languages, Platforms and Applications*, vol. 15 of *Multiagent Systems,
Artificial Societies, and Simulated Organizations*. Springer, Berlin.

[15] Bordini RH, Fisher M, Visser W and Wooldridge M 2004 Model checking rational
agents. *IEEE Intelligent Systems* **19**(5), 46 – 52.

[16] Bordini RH, Fisher M, Visser W and Wooldridge M 2006b Verifying multi-agent
programs by model checking. *Journal of Autonomous Agents and Multi-Agent Systems* **12**(2), 239 – 256.

[17] Bordini RH, Hübner JF and Tralamazza DM 2006c Using *Jason* to implement
a team of gold miners. In *Proceedings of the Seventh Workshop on Computational
Logic in Multi-Agent Systems (CLIMA VII), held with AAMAS 2006, 8 – 9 May,
Hakodate – Revised Selected and Invited Papers*, vol. 4371 of *LNAI*, pp. 304 – 313.
Springer, Berlin. (CLIMA Contest paper).

[18] Boyer RS and Moore JS (eds) 1981 *The Correctness Problem in Computer Science*.
Academic Press: London.

[19] Bratko I 1990 *Prolog Programming for Artificial Intelligence*, 2nd edn. Addison-
Wesley, Reading, MA.

[20] Bratman ME 1987 *Intention, Plans, and Practical Reason*. Harvard University Press,
Cambridge, MA.

[21] Bratman ME 1990 What is intention? In *Intentions in Communication* (ed. Cohen
PR, Morgan JL and Pollack ME), pp. 15 – 32. The MIT Press, Cambridge, MA.

[22] Bratman ME, Israel DJ and Pollack ME 1988 Plans and resource-bounded practical
reasoning. *Computational Intelligence* **4**, 349 – 355.

[23] Braubach L, Pokahr A, Lamersdorf W and Moldt D 2005 Goal representation for
BDI agent systems. In *Proceedings of the Second International Workshop on "Programming Multi-Agent Systems: Languages and Tools" (ProMAS 2004), held with the
Third International Joint Conference on Autonomous Agents and Multi-Agent Systems (AAMAS 2004), New York City, 20 July, Selected Revised and Invited Papers*

(ed. Bordini RH, Dastani M, Dix J and El Fallah-Seghrouchni A), vol. 3346 of *LNAI*. Springer, Berlin.

[24] Brooks RA 1999 *Cambrian Intelligence*. The MIT Press: Cambridge, MA.

[25] Busetta P, Howden N, Ronnquist R and Hodgson A 2000 Structuring BDI agents in functional clusters. In *Intelligent Agents VI — Proceedings of the Sixth International Workshop on Agent Theories, Architectures, and Languages, ATAL-99* (ed. Jennings N and Lespérance Y), vol. 1757 of *LNAI*, pp. 277 – 289. Springer, Berlin.

[26] Clark KL and McCabe FG 2004 Go! for multi-threaded deliberative agents. In *Declarative Agent Languages and Technologies, First International Workshop, DALT 2003, Melbourne, 15 July 2003, Revised Selected and Invited Papers* (ed. Leite JA, Omicini A, Sterling L and Torroni P), vol. 2990 of *LNAI*, pp. 54 – 75. Springer, Berlin.

[27] Clarke EM, Grumberg O and Peled DA 2000 *Model Checking*. The MIT Press: Cambridge, MA.

[28] Clocksin WF and Mellish C 1987 *Programming in Prolog*, 3rd edn. Springer, Berlin.

[29] Cohen PR and Levesque HJ 1990a Intention is choice with commitment. *Artificial Intelligence* **42**(3), 213 – 261.

[30] Cohen PR and Levesque HJ 1990b Rational interaction as the basis for communication. In *Intentions in Communication* (ed. Cohen PR, Morgan J and Pollack ME), pp. 221 – 256. The MIT Press, Cambridge, MA.

[31] Cohen PR and Perrault CR 1979 Elements of a plan based theory of speech acts. *Cognitive Science* **3**, 177 – 212.

[32] Dastani M and Gomez-Sanz J 2006 Programming multi-agent systems. *Knowledge Engineering Review* **20**(2), 151 – 164.

[33] Dastani M, Dix J and Novák P 2006 The second contest on multi-agent systems based on computational logic. In *Proceedings of the Seventh Workshop on Computational Logic in Multi-Agent Systems (CLIMA VII), held with AAMAS-2006, 8 – 9 May, Hakodate — Revised Selected and Invited Papers*, vol. 4371 of *LNAI*, pp. 266 – 283. Springer, Berlin. (CLIMA Contest paper).

[34] Dastani M, van Riemsdijk B, Hulstijn J, Dignum F and Meyer JJC 2004 Enacting and deacting roles in agent programming. In *Agent-Oriented Software Engineering V, 5th International Workshop, AOSE 2004, New York, 19 July 2004, Revised Selected Papers* (ed. Odell J, Giorgini P and Müller JP), vol. 3382 of *LNCS*, pp. 189 – 204. Springer, Berlin.

[35] Dastani M, van Riemsdijk MB and Meyer JJC 2005 Programming multi-agent systems in 3APL. In [14], Chapter 2, pp. 39 – 67.

[36] Davis R and Smith RG 1983 Negotiation as a metaphor for distributed problem solving. *Artificial Intelligence* **20**, 63 – 109.

[37] de Giacomo G, Lespérance Y and Levesque HJ 2000 ConGolog: a concurrent programming language based on the situation calculus. *Artificial Intelligence* **121**, 109 – 169.

[38] Dennett DC 1987 *The Intentional Stance*. The MIT Press, Cambridge, MA.

[39] d'Inverno M and Luck M 1998 Engineering AgentSpeak(L): a formal computational model. *Journal of Logic and Computation* **8**(3), 1 – 27.

[40] d'Inverno M, Kinny D, Luck M and Wooldridge M 1998 A formal specification of dMARS In *Intelligent Agents IV – Proceedings of the Fourth International Workshop on Agent Theories, Architectures, and Languages (ATAL-97), Providence, RI, 24 – 26 July 1997* (ed. Singh MP, Rao AS and Wooldridge M), vol. 1365 of *LNAI*, pp. 155 – 176. Springer, Berlin.

[41] d'Inverno M, Luck M, Georgeff MP, Kinny D and Wooldridge M 2004 The dMARS architecture: a specification of the distributed multi-agent reasoning system. *Autonomous Agents and Multi-Agent Systems* 9(1 – 2), 5 – 53.

[42] Dix J and Zhang Y 2005 IMPACT: a multi-agent framework with declarative semantics. In [14], Chapter 3, pp. 69 – 94.

[43] El Fallah Seghrouchni A and Suna A 2005 CLAIM and SyMPA: a programming environment for intelligent and mobile agents. In [14], Chapter 4, pp. 95 – 122.

[44] Finin T, Weber J, Wiederhold G, Genesereth M, Fritzson R, McKay D, McGuire J, Pelavin R, Shapiro S and Beck C 1993 Specification of the KQML agent communication language. DARPA Knowledge Sharing Initiative External Interfaces Working Group.

[45] FIPA 2003 FIPA contract net interaction protocol specification. Technical Report 00029, Foundation for Intelligent Physical Agents.

[46] Fisher M 2004 Temporal development methods for agent-based systems. *Autonomous Agents and Multi-Agent Systems* 10(1), 41 – 66.

[47] Fisher M, Bordini RH, Hirsch B and Torroni P 2007 Computational logics and agents: a road map of current technologies and future trends. *Computational Intelligence* 23(1), 61 – 91.

[48] Gamma E, Helm R, Johnson R and Vlissides J 1995 *Design Patterns: Elements of Reusable Object-Oriented Software*. Addison-Wesley, Reading, MA.

[49] Ghallab M, Nau D and Traverso P 2004 *Automated Planning: Theory and Practice*. Morgan Kaufmann, San Mateo, CA.

[50] Gilbert N and Conte R (eds) 1995 *Artificial Societies: the Computer Simulation of Social Life*. UCL Press, London.

[51] Gilbert N and Doran J (eds) 1994 *Simulating Society: the Computer Simulation of Social Phenomena*. UCL Press, London.

[52] Henderson-Sellers B and Giorgini P (eds) 2005 *Agent-Oriented Methodologies*. Idea Group, Hershey, PA.

[53] Holzmann GJ 2003 *The SPIN Model Checker: Primer and Reference Manual*. Addison-Wesley, Reading, MA.

[54] Huber M 1999 JAM: a BDI-theoretic mobile agent architecture *Proceedings of the Third International Conference on Autonomous Agents (Agents 99), Seattle, WA*, pp. 236 – 243.

[55] Hübner JF, Bordini RH and Wooldridge M 2007 Programming declarative goals using plan patterns. In *Proceedings of the Fourth International Workshop on Declarative Agent Languages and Technologies (DALT 2006), held with AAMAS 2006, 8 May, Hakodate – Selected, Revised and Invited Papers* (ed. Baldoni M and Endriss U), vol. 4327 of *Lecture Notes in Computer Science*, pp. 123 – 140. Springer, Berlin.

[56] Hübner JF, Sichman JS and Boissier O 2004 Using the $\mathcal{M}$oise+ for a cooperative framework of MAS reorganisation. In *Advances in Artificial Intelligence – SBIA*

2004, 17th Brazilian Symposium on Artificial Intelligence, São Luis, Maranhão, 29 September to 1 October, 2004, Proceedings (ed. Bazzan ALC and Labidi S), vol. 3171 of *LNAI*, pp. 506 – 515. Springer, Berlin.

[57] Huth M and Ryan M 2004 *Logic in Computer Science: Modelling and Reasoning about Systems*, 2nd edn. Cambridge University Press, Cambridge.

[58] Jennings NR 2000 On agent-based software engineering. *Artificial Intelligence* **117**, 277 – 296.

[59] Kinny D and Georgeff M 1991 Commitment and effectiveness of situated agents. In *Proceedings of the Twelfth International Joint Conference on Artificial Intelligence (IJCAI-91), Sydney*, pp. 82 – 88.

[60] Kumar S, Cohen PR and Huber MJ 2002 Direct execution of team specifications in STAPLE. In *Proceedings of the First International Joint Conference on Autonomous Agents and Multi-Agent Systems (AAMAS-2002), 15 – 19 July, Bologna* (ed. Castelfranchi C and Johnson WL), pp. 567 – 568. ACM Press, New York.

[61] Labrou Y and Finin T 1994 A semantics approach for KQML – a general purpose communication language for software agents *Proceedings of the Third International Conference on Information and Knowledge Management (CIKM'94)*. ACM Press, New York.

[62] Labrou Y and Finin T 1997 Semantics and conversations for an agent communication language. In *Proceedings of the Fifteenth International Joint Conference on Artificial Intelligence (IJCAI-97), Nagoya*, pp. 584 – 591.

[63] Leite JA 2003 *Evolving Knowledge Bases: Specification and Semantics* vol. 81 of *Frontiers in Artificial Intelligence and Applications, Dissertations in Artificial Intelligence*. IOS Press/Ohmsha, Amsterdam.

[64] Leite JA, Alferes JJ and Pereira LM 2002 $\mathcal{MINERVA}$ – a dynamic logic programming agent architecture. In *Intelligent Agents VIII – Proceedings of the Eighth International Workshop on Agent Theories, Architectures, and Languages (ATAL-2001), 1 – 3 August 2001, Seattle, WA* (ed. Meyer JJ and Tambe M), vol. 2333 of *LNAI*, pp. 141 – 157. Springer, Berlin.

[65] Lifschitz V 1986 On the semantics of STRIPS. In *Reasoning About Actions & Plans – Proceedings of the 1986 Workshop* (ed. Georgeff MP and Lansky AL), pp. 1 – 10. Morgan Kaufmann, San Mateo, CA.

[66] Lloyd JW 1987 *Foundations of Logic Programming*, 2nd edn. Springer, Berlin.

[67] Mařík V and Vrba P 2004 Developing agents for manufacturing at Rockwell automation. *AgentLink News* **16**, 9 – 11.

[68] Mascardi V, Martelli M and Sterling L 2004 Logic-based specification languages for intelligent software agents. *Theory and Practice of Logic Programming* **4**(4), 429 – 494.

[69] Mayfield J, Labrou Y and Finin T 1996 Evaluation of KQML as an agent communication language. In *Intelligent Agents II – Proceedings of the Second International Workshop on Agent Theories, Architectures, and Languages (ATAL'95), held as part of IJCAI'95, Montréal, August 1995* (ed. Wooldridge M, Müller JP and Tambe M), vol. 1037 of *LNAI*, pp. 347 – 360, Springer, Berlin.

[70] Moreira ÁF, Vieira R and Bordini RH 2004 Extending the operational semantics of a BDI agent-oriented programming language for introducing speech-act based communication. In *Declarative Agent Languages and Technologies, Proceedings of the First*

International Workshop (DALT-03), held with AAMAS-03, 15 July 2003, Melbourne, (Revised Selected and Invited Papers) (ed. Leite J, Omicini A, Sterling L and Torroni P), vol. 2990 of *LNAI*, pp. 135 – 154. Springer, Berlin.

[71] Moreira ÁF, Vieira R, Bordini RH and Hübner JF 2006 Agent-oriented programming with underlying ontological reasoning. In *Proceedings of the Third International Workshop on Declarative Agent Languages and Technologies (DALT-05), held with AAMAS-05, 25 July, Utrecht* (ed. Baldoni M, Endriss U, Omicini A and Torroni P), vol. 3904 of *Lecture Notes in Computer Science*, pp. 155 – 170. Springer, Berlin.

[72] Morley D and Myers KL 2004 The SPARK agent framework *3rd International Joint Conference on Autonomous Agents and Multiagent Systems (AAMAS 2004), 19 – 23 August 2004, New York*, pp. 714 – 721. IEEE Computer Society, New York.

[73] Okuyama FY, Bordini RH and da Rocha Costa AC 2005 ELMS: an environment description language for multi-agent simulations In *Environments for Multiagent Systems, State-of-the-art and Research Challenges. Proceedings of the First International Workshop on Environments for Multiagent Systems (E4MAS), held with AAMAS-04, 19 July* (ed. Weyns D, van Dyke Parunak H, Michel F, Holvoet T and Ferber J), vol. 3374 of *LNAI*, pp. 91 – 108. Springer, Berlin.

[74] Okuyama FY, Bordini RH and da Rocha Costa AC 2006 Spatially distributed normative objects. In *Proceedings of the Workshop on Coordination, Organization, Institutions and Norms in Agent Systems (COIN), held with ECAI 2006, 28 August, Riva del Garda* vol. 4386 of *LNAI*, pp. 133 – 146. Springer, Berlin.

[75] Padgham L and Winikoff M 2004 *Developing Intelligent Agent Systems: a Practical Guide*. Wiley, Chichester.

[76] Plotkin GD 1981 A structural approach to operational semantics. Technical report, Computer Science Department, Aarhus University, Aarhus, Denmark.

[77] Pnueli A 1986 Specification and development of reactive systems. In *Information Processing 86*, pp. 845 – 858. Elsevier Science, Amsterdam.

[78] Pokahr A, Braubach L and Lamersdorf W 2005 Jadex: a BDI reasoning engine. In [14], Chapter 6, pp. 149 – 174.

[79] Prietula M, Carley K and Gasser L (eds) 1998 *Simulating Organizations: Computational Models of Institutions and Groups*. AAAI Press/MIT Press, Menlo Park, CA.

[80] Rao AS 1996 AgentSpeak(L): BDI agents speak out in a logical computable language. In *Proceedings of the Seventh Workshop on Modelling Autonomous Agents in a Multi-Agent World (MAAMAW'96), 22 – 25 January, Eindhoven* (ed. Van de Velde W and Perram J), vol. 1038 of *LNAI*, pp. 42 – 55. Springer, London.

[81] Rao AS and Georgeff MP 1991 Modeling rational agents within a BDI-architecture. In *Proceedings of the 2nd International Conference on Principles of Knowledge Representation and Reasoning (KR'91)* (ed. Allen J, Fikes R and Sandewall E), pp. 473 – 484. Morgan Kaufmann, San Mateo, CA, USA.

[82] Russell S and Norvig P 1999 *Artificial Intelligence: a Modern Approach*, 3rd edn. Prentice Hall, Englewood Cliffs, NJ.

[83] Sardiña S, de Silva L and Padgham L 2006 Hierarchical planning in BDI agent programming languages: a formal approach. In *5th International Joint Conference on*

Autonomous Agents and Multiagent Systems (AAMAS 2006), Hakodate, 8 – 12 May 2006 (ed. Nakashima H, Wellman MP, Weiss G and Stone P), pp. 1001 – 1008. ACM Press, New York.

[84] Schut M and Wooldridge M 2001 The control of reasoning in resource-bounded agents. *The Knowledge Engineering Review* **16**(3), 215 – 240.

[85] Schut MC, Wooldridge M and Parsons S 2004 The theory and practice of intention reconsideration. *Journal of Experimental and Theoretical Artificial Intelligence* **16**(4), 261 – 293.

[86] Searle JR 1969 *Speech Acts: an Essay in the Philosophy of Language*. Cambridge University Press, Cambridge.

[87] Shoham Y 1993 Agent-oriented programming. *Artificial Intelligence* **60**(1), 51 – 92.

[88] Singh M and Huhns MN 2005 *Service-Oriented Computing: Semantics, Processes, Agents*. Wiley, Chichester.

[89] Singh MP 1994 *Multiagent Systems – a Theoretic Framework for Intentions, Know-How, and Communications*, vol. 799 of *LNAI*. Springer, Berlin.

[90] Smith RG 1980 The contract net protocol: high-level communication and control in a distributed problem solver. *IEEE Transactions on Computers* **29**(12), 1104 – 1113.

[91] Staab S and Studer R (eds) 2004 *Handbook on Ontologies*. International Handbooks on Information Systems. Springer, Berlin.

[92] Sterling L and Shapiro EY 1994 *The Art of Prolog – Advanced Programming Techniques*, 2nd edn. MIT Press, Cambridge, MA.

[93] Sun R (ed.) 2005 *Cognition and Multi-Agent Interaction: From Cognitive Modeling to Social Simulation*. Cambridge University Press, Cambridge.

[94] Tambe M 1997 Agent architectures for flexible, practical teamwork. *AAAI/IAAI*, pp. 22 – 28.

[95] Thangarajah J 2004 *Managing the Concurrent Execution of Goals in Intelligent Agents*. PhD thesis School of Computer Science and Information Technology, RMIT University, Melbourne.

[96] Thangarajah J, Padgham L and Winikoff M 2003 Detecting and avoiding interference between goals in intelligent agents. In *Proceedings of 18th International Joint Conference on Artificial Intelligence (IJCAI)*, pp. 721 – 726. Morgan Kaufmann, Acapulco.

[97] Toni F 2006 Multi-agent systems in computational logic: challenges and outcomes of the SOCS project. In *Computational Logic in Multi-Agent Systems, 6th International Workshop, CLIMA VI, London, UK, 27 – 29 June, 2005, Revised Selected and Invited Papers* (ed. Toni F and Torroni P), vol. 3900 of *Lecture Notes in Computer Science*, pp. 420 – 426. Springer, Berlin.

[98] van Riemsdijk B, Dastani M and Meyer JJC 2005 Semantics of declarative goals in agent programming. In *4rd International Joint Conference on Autonomous Agents and Multiagent Systems (AAMAS 2005), 25 – 29 July 2005, Utrecht* (ed. Dignum F, Dignum V, Koenig S, Kraus S, Singh MP and Wooldridge M), pp. 133 – 140. ACM Press, New York.

[99] van Riemsdijk MB 2006 *Cognitive Agent Programming: a Semantic Approach*. PhD thesis, Utrecht University, Utrecht.

[100] Vieira R, Moreira A, Wooldridge M and Bordini RH 2007 On the formal semantics of speech-act based communication in an agent-oriented programming language. *Journal of Artificial Intelligence Research* **29**, 221 – 267.

[101] Winikoff M 2005 JACK™ intelligent agents: an industrial strength platform. In [14], Chapter 7, pp. 175 – 193.

[102] Winikoff M, Padgham L, Harland J and Thangarajah J 2002 Declarative and procedural goals in intelligent agent systems. In *Proceedings of the Eighth International Conference on Principles of Knowledge Representation and Reasoning (KR2002), 22 – 25 April, Toulouse*, pp. 470 – 481.

[103] Wooldridge M 2002 *An Introduction to MultiAgent Systems*. Wiley, Chichester.

[104] Wooldridge M and Jennings NR 1995 Intelligent agents: theory and practice. *The Knowledge Engineering Review* **10**(2), 115 – 152.

Index

Printed and bound by CPI Group (UK) Ltd, Croydon, CR0 4YY

27/10/2024

14580148-0003